Digital Transformation of Oil and Gas

Intelligent Fields and Workflows

Sami Alnuaim

WILEY-VCH

Author

Dr. Sami Alnuaim
College of Petroleum Engineering & Geosciences
King Fahd University of Petroleum & Minerals
Dhahran 31311
Saudi Arabia

Cover Design: Wiley
Cover Image: Issarawat Tattong/Getty Images

Typesetting: Lumina Datamatics, Madurai, India

Library of Congress Card No.: applied for

British Library Cataloguing-in-Publication Data
A catalogue record for this book is available from the British Library.

Bibliographic information published by the Deutsche Nationalbibliothek The Deutsche Nationalbibliothek lists this publication in the Deutsche Nationalbibliografie; detailed bibliographic data are available on the Internet at <http://dnb.d-nb.de>.

Print ISBN: 9783527357161
ePDF ISBN: 9783527858750
ePub ISBN: 9783527858743
oBook ISBN: 9783527858767

Printed and bound by
CPI Group (UK) Ltd, Croydon, CR0 4YY

C978352735161_190826

This work is dedicated to the following people, who contributed to my learning and success:

- My father and mother, who taught me the skills of excellence, continuous learning, and ethics.
- My wife and family, who supported me and encouraged me to write this book.
- Saudi Aramco, the greatest energy company for my education sponsorships (B.S., M.S., and Ph.D.), for the technological exposure, business ethics, excellence, and values I learned from.
- Last, King Fahd University of Petroleum and Minerals (KFUPM), which availed me the golden opportunity to develop this course and technological domain and teach it, including enriching the research side, ending up with this book.

My long personal scientific journey has always been guided by these two wonderful verses in the Holy Quran calling for continuous excellence in Scientific Pursuit:

﴿إِنَّ فِي خَلْقِ السَّمَاوَاتِ وَالْأَرْضِ وَاخْتِلَافِ اللَّيْلِ وَالنَّهَارِ وَالْفُلْكِ الَّتِي تَجْرِي فِي الْبَحْرِ بِمَا
يَنفَعُ النَّاسَ وَمَا أَنزَلَ اللَّهُ مِنَ السَّمَاءِ مِن مَّاءٍ فَأَحْيَا بِهِ الْأَرْضَ بَعْدَ مَوْتِهَا وَبَثَّ فِيهَا مِن كُلِّ
دَابَّةٍ وَتَصْرِيفِ الرِّيَاحِ وَالسَّحَابِ الْمُسَخَّرِ بَيْنَ السَّمَاءِ وَالْأَرْضِ لَآيَاتٍ لِّقَوْمٍ يَعْقِلُونَ﴾

القران الكريم (البقرة أية 164)

"Verily! In the creation of the heavens and the earth, and in the alternation of night and day, and the ships which sail through the sea with that which is of use to mankind, and the water (rain) which Allah sends down from the sky and makes the earth alive therewith after its death, and the moving (living) creatures of all kinds that He has scattered therein, and in the veering of winds and clouds which are held between the sky and the earth, are indeed Ayat (proofs, evidences, signs, etc.) for people of understanding."

Holly Quran, Chapter 2—Verse 164

﴿الَّذِينَ يَذْكُرُونَ اللَّهَ قِيَامًا وَقُعُودًا وَعَلَىٰ جُنُوبِهِمْ وَيَتَفَكَّرُونَ فِي خَلْقِ السَّمَاوَاتِ وَالْأَرْضِ رَبَّنَا
مَا خَلَقْتَ هَٰذَا بَاطِلًا سُبْحَانَكَ فَقِنَا عَذَابَ النَّارِ﴾

القران الكريم (العمران أية 191)

"Those who remember Allah (always, and in prayers) standing, sitting, and lying down on their sides, and think deeply about the creation of the heavens and the earth, (saying): 'Our Lord! You have not created (all) this without purpose, glory to You! (Exalted be You above all that they associate with You as partners). Give us salvation from the torment of the Fire.'"

Holly Quran, Chapter 3—Verse 191

To those advancing the future of energy that has brought happiness and prosperity to all humanity.

Contents

About the Author

Dr. Sami Alnuaim

An international energy expert, consultant, academic, writer, and researcher in the fields of Digital Oil Fields and Energy Sustainability. He has over 35 years of administrative, technical, and academic experience, the majority of which has been with Saudi Aramco and King Fahd University of Petroleum and Minerals in the digital oil and gas fields and global energy sustainability domains. Dr. Alnuaim has served in the following roles:

- The 2019 President of the Society of Petroleum Engineers (SPE.org), the largest petroleum engineering professional society.
- Adjunct Professor at King Fahd University of Petroleum and Minerals, Saudi Arabia.
- Member involved in teaching Digital Oil Field and Energy Sustainability to graduate and undergraduate students.
- Member of the International Advisory Board of the University of Texas, Austin, United States.
- Member of the Industrial Advisory Board of the College of Engineering at King Faisal University, Saudi Arabia.
- Past member of the Board of Directors of the Society of Petroleum Engineers (SPE.org).
- Recipient of the Society of Petroleum Engineers' International 2011 Excellence Award for Service to the Petroleum Engineering Sector.
- Recipient of the 2023 USCLA Danish Energy Sustainability Excellence Award, Los Angles California.
- International keynote speaker in oil, gas, and energy conferences.

Preface and Acknowledgments

The global energy landscape has been undergoing a profound digital and intelligent transformation. Traditional oil and gas operations, once dependent on manual labor, intuition, and segmented systems, are now being fundamentally reshaped by digital technology and intelligent information techno-innovation at an exponentially increased pace. The oil and gas industry, which has been labeled as a slow and low-tech industry in the last century, has gone through evolution, rising to the concept of smart and intelligent fields since the beginning of the new millennium. This revolutionary approach has effectively integrated the unmatched solid physical infrastructure of oil and gas assets with basic engineering and IT fundamentals through advanced technologies such as digitization, robotics, automation, integration, artificial intelligence (AI), machine learning (ML), the Internet of Things (IoT), edge computing, digital twin, and real-time data analytics.

This book, **Digital Transformation of Oil and Gas: Intelligent Fields and Workflows**, explores this ongoing and exciting oil and gas industry transformation journey, examining how digital intelligence is reshaping the industry in all its domains, exploration, drilling, fields development, production, reservoir management, and abandonment of hydrocarbon resources. The notion of the smart field is not merely about automation or the use of high-tech gadgets; rather, it is about creating a fully integrated ecosystem, where real-time data-driven insights articulated by skilled engineers and data scientists lead to better decision-making, increased efficiency, enhanced safety, reduced environmental impact, and, ultimately, improved corporate profitability and sustainability performances.

The oil and gas industry is facing mounting challenges: increasing demand, volatile commodity prices, rising operational costs, regulatory pressures, global climate challenge imperatives, and increasing demand for cleaner and more efficient energy sources with net-zero emissions challenging targets and near-zero environmental impact. Traditional methods, while effective in the past, are no longer sufficient to address the complexity and urgency of modern energy needs. This exciting journey is one of the major factors that helped the industry to effectively control the development cost and hence manage the energy cost from reaching extremely high prices that could negatively impact the development of the challenging assets such as shale/unconventional oil and gas and preserving the unprecedented global economic growth, sustainable social development, human development index, and relatively low energy poverty indices.

With the evolution of industries of other energy sources and to remain viable and competitive, oil and gas companies had to adopt efficient, predictive, and resilient smart/intelligent modes of operation. The move toward Smarter and Intelligent Fields is not just a technological upgrade but rather a strategic industry imperative and necessity, without which the industry will not be able to compete and survive.

This book is designed to serve as a comprehensive guide for students, professionals, engineers, executives, policymakers, and academics interested in the future of the oil and gas industry. Each chapter describes a critical aspect of the smart and intelligent digital oil and gas fields, including the following:

- Chapter 1: Evolution of Digital Oil Fields (DOFs)—Historical Context and Technological Advancements
- Chapter 2: DOF Strategic Objectives
- Chapter 3: DOF Infrastructure and Asset Management
- Chapter 4: DOF Financial Aspects and Business Planning
- Chapter 5: DOF Organizational Change and Workforce Transformation
- Chapter 6: DOF Cybersecurity, Business Continuity, and Disaster Recovery
- Chapter 7: Enterprise DOF Workflows—Field Operation and Production Engineering
- Chapter 8: Enterprise DOF Workflows—Reservoir Monitoring and Well Testing
- Chapter 9: Enterprise Strategic DOF Workflows
- Chapter 10: DOF and Corporate Management
- Chapter 11: DOF Future: Toward Fully Autonomous and Sustainable DOF
- Appendix A: Typical Large-scale DOF Implementations
- Appendix B: DOF AI Case Study: Electrical Submersible Pump (ESP) Failure Prediction
- Appendix C: DOF Readiness Index—A New Proposed Framework for Smart Field Transformation

The book summarizes my 35 years of working experience with the greatest energy company, Saudi Aramco, spent on several upstream organizations, including Reservoir Management, Production Engineering, and Upstream Research and Development and ending with the DOF hosting organization. Hence, many aspects of this book are derived from my personal long DOF experience, my personal understanding of the maturity of this technology, and its significant impact on the industry's technical-leadership, profitability, efficiency, and sustainable performance. In addition, the book content was complemented by additional information that was extracted from several cited technical references. Chapters are written in a generic language style, structured uniquely based on my understanding of the DOF concept and linguistically improved using AI models (ChatGPT Pro V5.2). All explanations, descriptions, definitions, historical summaries, and technical concepts are written in my own academic/technical wording unless otherwise quoted. Plots, diagrams, and illustrations used in this book are either reproduced from sources that are mentioned, quoted/cited, or AI-generated/self-customized (most of the illustrations).

Furthermore, and for the first time, the comprehensive and rich class notes and PowerPoint presentations of my KFUPM smart fields and reservoir monitoring graduate course—that I have been teaching since 2020—have been used to train an AI (ChatGPT Pro V5.2) model to develop parts of the chapter's narratives, supported by several examples and case studies found in public domain sources. The book was then thoroughly and technically checked for better quality assurance by me and an industry DOF expert (Tofig Al-Dhubaib, retired Saudi Aramco DOF senior consultant).

To avoid any commercial content, all DOF equipment specifications mentioned in this book were collected from different public domain sources. Most of these illustrations/pictures of this equipment and other relevant technologies were AI generated (unless mentioned) and customized in generic forms, regardless of the manufacturer. Permission was obtained from the publisher(s) where I used a few figures as is (in Chapters 3 and 4). Figures, tables, and information from other sources that have been rewritten and reconstructed have been referenced as per the best-in-class copyright practices. Hence, most of them do not reproduce copyrighted text verbatim from any identifiable published source or reference, unless specifically mentioned. The citation style will be in either combined groups of references or only the most appropriate reference using the Vancouver style.

Finally, I would like to end the Preface section with this standard disclaimer:

The views and opinions expressed in this book are solely those of the author and do not reflect the policies or positions of any entity the author has worked for in the past, or any affiliated or associated entities, each of whom disclaim any responsibility or liability for any violation of intellectual property or secret rights and waive no rights, privileges, immunities, or defenses under applicable law. Furthermore, the author has absolutely no responsibility or liability as a result of any use of the data, figures, or opinions expressed in this book.

Let us now begin our journey into this exciting, intelligent frontier.

1

Evolution of Digital Oil Fields (DOFs)

Historical Context and Technological Advancements

From Conventional Analog to Smart Wells and Autonomous Operations, Maximizing Recovery, Profitability, Health, Safety, Environment, and Sustainability Performance (everythingpossible/Adobe Stock Photos).

The oil and gas industry has long been defined by its engineering skills, capital intensity, and ability to operate in some of the world's harshest environments, such as deserts, deep seas, and arctic remote geographical frontiers. However, until recently, much of its core infrastructure remained rooted in legacy analog manual systems—mechanical gauges, manual inspections, and siloed static data. This chapter introduces the concept of smart wells/fields and highlights the key technological milestones, paradigm shifts, and foundational innovations that have shaped the evolution of the digital oil and gas fields over the last few decades.

1.1 Defining Smart and Intelligent Wells and Fields

1.1.1 Smart Wells

In the 1990s, the concept of smart well completion was explored and tested by several national and international oil companies (NOCs/IOCs) to address complex wells' production challenges, such as water coning, fractures (super-K), uneven sweep efficiency, and low recovery factors. At the end of the 20th century, a number of complete sets of smart wells' completion products have been developed by major oil services companies (OSCs) and put into production in the applications of several fields, especially by major NOCs and IOCs (e.g., Saudi Aramco). After decades of development, large-scale oil and gas production equipment enterprises led by major OSCs have further improved the technology of intelligent wells from the aspects of electrification, electronics, remote communication, hierarchical control, and smart applications. Hence, smart wells can be defined as:

- Wells that are equipped with preinstalled high-tech devices in the wellbore that help in protecting the well integrity, optimizing well productivity, and monitoring and controlling the well from the surface, electrically or hydraulically.
- Wells that can self-control, collect and analyze data, and potentially make decisions that would increase recovery, improve safety, and reduce operating costs.

Smart wells have the following strategic and operational objectives:

- Optimize production/injection and hence improve hydrocarbon production and ultimate recovery.
- Ensure well integrity and hence reduce well interventions and maintenance costs.
- Detect surface and downhole anomalies, leak, emissions, and risk factors and hence improve health, safety, and environment (HSE).

Smart well technologies have become a necessity to enhance the oil and gas companies' profitability, HSE, and sustainability performances. Today, smart wells drive most of the upstream development projects to be cost-effective, increasing their projects' net present value (NPV). This can be achieved by reducing the number of drilled wells, workovers, and intervention operations, hence reducing the overall project development cost ($/barrel) and environmental footprint [1–10].

1.1.2 Intelligent Oil and Gas Fields

At its core, a smart field (sometimes referred to as a digital oil field [DOF] or intelligent field) integrates real-time data from various sources—smart wells, pipelines, reservoirs, and surface and subsurface facilities—and processes using advanced analytics and artificial intelligence (AI) to protect and optimize operations. These fields use a continuous loop of data acquisition, communication, analysis, and smart proactive safety and engineering actions. This enables proactive predictive maintenance, real-time production optimization, automated drilling decisions, and even autonomous field assets management. An intelligent field goes a step further, introducing adaptive systems that learn from past data, improve over time, and make autonomous recommendations or decisions. The difference lies in cognition: whereas a smart field collects and analyses, an intelligent field anticipates and evolves to a degree where intelligent DOFs have become a necessity to address the following oil and gas industry strategic objectives:

- Increase **efficiency** and **reliability** of upstream operations (oil and gas energy supply).
- Increase **productivity**, **reserves**, and **ultimate recovery** (oil and gas energy security).
- Reduce **development cost** ($/barrel).
- Enhance **safety**, **environmental protection**, and **sustainability** (HSE and sustainability performance).

DOF key features (must-to-have):

- **Real-time Monitoring**: Use of sensors (e.g., downhole, surface, and subsea) to collect continuous data on pressure, temperature, flow rates, and so on.
- **Local, Regional, and Wide Networks**: Networks necessary for connectivity and real-time data transfer.
- **Integrated Operations Centers (IOCs)**: Centralized hubs that analyze data, allowing remote monitoring/control of field equipment and operations.
- **Advanced Modeling and Simulation**: Well/reservoir models are dynamically updated with real-time data to improve forecasts and production planning.
- **Automation and Control Systems**: Technologies like supervisory control and data acquisition (SCADA) and distributed control systems (DCS) enable real-time decision-making and operational automation.
- **Collaborative Environments**: Multidisciplinary teams (e.g., geoscientists, engineers, data scientists) work together using shared data and advanced smart tools to make better decisions.
- **Engineering and Operational Workflows**: A set of multidomain algorithms to use the real-time data, providing predictive, preventive, and proactive decision-making environments.

1.1.3 Brown Conventional Oil Fields

A conventional non-smart oil field is a traditional oil and gas field that relies on standard exploration, drilling, and production methods, without the integration of real-time monitoring, advanced analytics, or automated control systems. These fields typically use manual operations, periodic data acquisition, and legacy infrastructure. Decision-making is generally reactive, based on overdue historical performance rather than real-time insights.

Brown fields' key characteristics:

- **Manual or Periodic Data Collection**: Data is collected at intervals using surface and downhole gauges, not continuously.
- **Limited Automation**: Most operational controls (e.g., valve adjustments, well testing) are done manually or semimanually.
- **Standard Recovery Techniques**: Production methods rely on natural reservoir pressure, artificial lift, and basic water flooding, with minimal use of enhanced or smart technologies.
- **Isolated Systems**: Analog systems and equipment and operational units function independently with minimal integration.
- **Delayed Decision-making**: Operational decisions are based on reports and delayed feedback rather than real-time optimization. Depending on each operator's data collection periodic strategy, serious well productivity, safety, and environmental problems could take months or years to be discovered.

This contrasts with smart oil fields, which are digitally integrated, data-driven, and optimized in near real time.

The smart and intelligent fields journey was not a sudden leap but rather the result of decades of technological progression and the industry's commitment toward technology, excellence, and innovation led by major NOCs, IOCs, and OSCs.

1.2 The Early Days: Conventional Wells and Analog Fields' Operations

In the early and mid-20th century, oil field operations were largely mechanical. Measurements were taken using analog instruments, and data were recorded manually in logbooks. Communication between field assets and control rooms was limited, and real-time decision-making was practically nonexistent [1–10].

Key negative characteristics of this era included:

- **Manual Oil and Gas Well Monitoring**: Operators physically visited wells to record flow rates and pressures using analog gauges with a relatively high error margin.
- **Reactive Maintenance**: Equipment failures often resulted in unplanned downtime, leading to costly repairs, production reduction, increased development cost, and potential safety and environmental hazards.
- **Reactive Well Subsurface Integrity**: Field issues took very long lagging periods to be identified due to the limited manual multiyear periodic well tests and integrity surveys. Example of well's subsurface temperature surveys that could detect subsurface leaks were usually run manually every several years.
- **Data delays**: Weeks, months, and maybe years could pass between data acquisitions due to remote areas and harsh environments, especially in a harsh deep-sea environment.

- **Limited integration**: Reservoir, drilling, and production data were typically handled manually by separate departments or vendors and kept in disintegrated physical places, each with its own programs, practices, and data structure.

Despite these constraints, oil and gas companies thrived due to relatively easy access to abundant reserves and high margins. This was acceptable due to the low global energy demand during these early days, which has changed dramatically afterward.

1.3 The 1980s–1990s: Digital Journey Starts

The first digital revolution in the oil and gas industry began in the 1980s and 1990s with the introduction of computer-based technologies. Supervisory control and data acquisition (SCADA) systems allowed for basic remote monitoring and control of assets, while basic desktop software enabled more advanced reservoir modeling and seismic interpretation. Having access to supercomputers' mainframe provided the industry with more opportunities to digitize and model the surface and subsurface oil and gas assets with high-performance computing (HPC) software and hardware, enabling the industry to advance its modeling capabilities [1–10].

Key digital information technology (IT) revolution developments:

- **SCADA Systems**: Enabled remote surveillance of surface equipment such as oil and gas wells and offshore platforms, reducing the need for on-site visits, especially in harsh and remote areas such as offshore locations.
- **Distributed Control System (DCS)**: DCS systems enabled the industry to control intensive and complex industrial processes to increase their safety, cost-effectiveness, and reliability. A DCS is a combination of sets of advanced algorithms that manage mechanical and electrical devices that regulate other devices or systems using control programs, enabling remote surveillance of surface equipment in industrial plants, reducing the need for on-site visits.
- **Geographic Information Systems (GISs)**: GIS provided spatial analysis capabilities for exploration and development.
- **Digital 3D Seismic Processing**: Enhanced subsurface imaging accuracy, improving exploration success rates and reducing the number of exploratory wells and hence reducing the oil and gas field's development cost and its relevant environmental impact.
- **Corporate Relational Databases**: Structured corporate databases such as Oracle have enabled very effective storage of time-series operational data in relational tables supported with structured query languages (SQL), enabling easier database consolidation, management, auditing, security, and access control.

These tools laid the groundwork for greater data integration and set the stage for DOF and future automation.

1.4 The 2000s: Birth of the Digital Oil and Gas Fields

The term digital oil field (DOF) gained prominence in the early 2000s, as industry leaders began to conceptualize a more interconnected approach to field development and operations. A DOF aims to digitize, integrate, and automate key aspects of the hydrocarbon lifecycle using centralized data platforms and advanced analytics [1–10].

Core technologies and principles of DOF during this period:

- **Advances in IT:** The evolution of IT, including satellite and cellular networks, enabled the integration of previously mentioned technologies with real-time data streams from the fields to the computer centers to smartphones.
- **Internet of Things (IoT) and Surface/Subsurface Sensor Networks**: IoT provided the infrastructure and capability for real-time monitoring and control of equipment, wells, reservoirs, fields, and industrial plants.
- **Digital Twins**: Virtual replicas of physical assets or systems that simulate performance and predict failures are used either during the lack/interruption of real-time data streams or to verify the real-time data stream outcomes.
- **Advanced Data Acquisition Digital Equipment**: Real-time data acquisition through downhole sensors and surface instrumentation has been introduced and used, connecting smart well completion technologies such as inflow control devices, inflow control valves, permanent downhole measuring system, distributed temperature system, electric submersible pump, and multiphase flow measurement.
- **Integrated Operations Centers**: These advanced centers, equipped with high-performance computing, supported by AI/machine learning (ML) and manned by multidisciplinary skilled staff, helped the industry harvest DOF outcomes and solve very complex problems to achieve the DOF's expected objectives.
- **Automated Workflows**: Automated workflows supported by advanced analytics, AI, ML, and HPC in drilling, production, reservoir management, and maintenance, help the industry to scale up the DOF's expected objectives.

1.5 The 2010s: Digital Maturity and AI Emergence

The 2010s marked a period of digital maturity, driven by a convergence of big data, cloud computing, and the mature adoption of AI and ML. Operators began implementing data lakes and cloud-based analytics platforms to process vast volumes of structured and unstructured data [1–10].

Technological milestones:

- **Big Data and Cloud Computing**: Allow for scalable storage and very fast and efficient processing of vast volumes of structured and unstructured data.

- **AI and ML**: Enable pattern recognition, early proactive anomaly detection, and advanced predictive modeling.
- **Edge Computing**: Facilitates immediate data processing at or near the data source, crucial for remote or offshore operations.
- **Private and Public Cloud**: Private and public cloud computing adoption (e.g., Microsoft Azure and Amazon Web Services for energy).
- **Mobility and Smartphone Technology**: Mobile and remote monitoring via smartphones and tablets add mobility to the intelligent systems.
- **Advances in AI and ML**: AI and ML technologies significantly reduce exploration risk, enhance drilling efficiency, reduce drilling troubles, optimize well placement, improve hydraulic fracturing, optimize production through advanced smart well completions, and reduce development cost through predictive maintenance and real-time reservoir management.
- **AI, ML, and Digital Twins Integration**: The integration and quick development/deployment of these advanced technologies helped the industry to effectively simulate and optimize most of its critical physical surface and subsurface systems in real-time at the well, network, and field levels.

During this decade, upstream surface and subsurface data have grown exponentially in types, variety, and volume (Figure 1.1), fitting the definition of big data. In addition, the role of the petroleum engineer started to shift from hands-on problem-solving to managing and interpreting real-time digital data, patterns, and complex engineering workflows. Most, if not all, the manual sequential, disintegrated reactive systems have been transformed to parallel, fully integrated predictive systems and workflows, freeing most of the engineers' time to focus on strategies and high value-added opportunities.

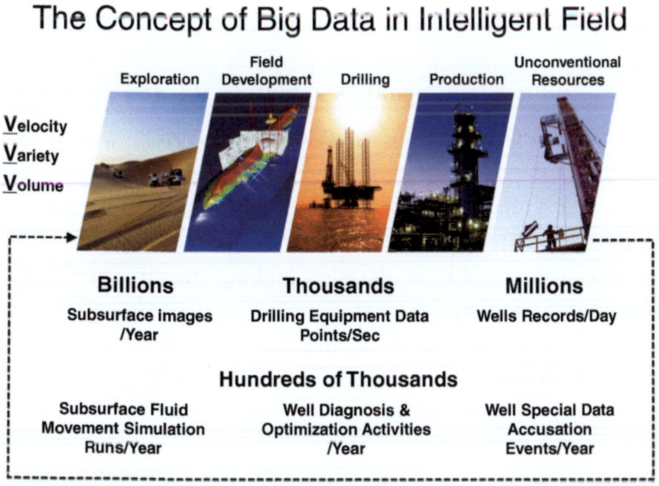

Figure 1.1 Generic upstream surface and subsurface data types (generated with AI using OpenAI).

Table 1.1 Comparison of annual data collected from brown and smart intelligent fields.

Conventional Field	Intelligent Field
Example of 100 wells field: 15% Key monitoring wells: Monthly rates tests Quarterly pressure test 85% Regular wells: Monthly rates tests Annual pressure tests **Annual readings rate & pressure = 1,345**	Example of 100 wells field: 100% monitored wells: Every 5 seconds rates reading Every 5 seconds pressure reading **Annual readings rate & pressure = 1.2 B** **Monthly readings rate & pressure = 100 M** **Daily readings rate & pressure = 3.5 M**

By doing a simple basic comparison between two similar fields, one brown and one intelligent, for only two parameters such as well oil rate and bottom-hole pressure, we can see (Table 1.1) a huge difference of 900,000 folds of annual collected data for the DOF compared to the brown field.

Any average intelligent field will have more than 30–50 operational real-time measured parameters per well, which makes this increase in data folds much bigger, where conventional field management may not be applicable. To deal with this massive data in a conventional way, you need an army of engineers. Using DOF, this task is fully integrated and automated, where few engineers are needed.

1.6 The 2020s: Smart and Intelligent Fields Boom

Today, the industry is transitioning from merely digital to a smart industry, increasingly partially/fully automated, intelligent, unmanned oil and gas fields. This era is characterized by robotics, automation, cognition, autonomy, AI, and predictive algorithms [1–10].

Key capabilities:

- Self-optimizing wells that adjust production settings autonomously based on real-time reservoir data supported by AI.
- AI-driven drilling platforms capable of decision-making during operations with minimum human intervention.
- Autonomous robots and drones used for data collection, inspection, and maintenance, especially in harsh areas.
- Full-field optimization algorithms, integrating subsurface, surface, and key financial and energy market parameters.
- Closed-loop control systems that act on predictive insights without human intervention.

What distinguishes intelligent fields in this decade is their capacity to learn, adapt, and act with minimal human input. Decisions are no longer just data-informed—they are data-driven and often data-execution.

1.7 Challenges Along the Way

While the DOF evolution is impressive, it has not been without obstacles:

- Workforce skill gaps in digital technologies and change resistance from traditional engineering cultures.
- Data fragmentation across legacy systems and digitization maturity, especially in small-medium independent operators.
- Cybersecurity vulnerabilities in interconnected assets.
- Integration complexity across vendors, equipment, and data protocols.

These challenges highlight the need for holistic digitization maturity, corporate transformation—technological, organizational, and cultural—and corporate cybersecurity strategy [1–10].

Despite these challenges and as the industry moves forward, the smart and intelligent oil and gas field is becoming the norm, not the exception. The next DOF frontier includes, but not limited to, the following technologies:

- AI-driven autonomous fields that operate with minimal human presence.
- Decentralized energy systems with real-time optimization across networks.
- Blockchain for secure data exchange between partners and regulators.
- Quantum computing to solve complex reservoir simulations, especially for giant reservoirs and fields such as Saudi Arabian reservoirs and fields.
- Sustainable intelligence that balances operational/financial performance with environmental and sustainable social development responsibilities, including emission monitoring and control toward the targets of net-zero emissions.
- The full integration of oil, gas, and renewable energy sources, driven by advanced AI/ML global energy demand models with real-time smart optimization algorithms.

The evolution from analog wells to autonomous fields marks one of the most significant transformations in the history of oil and gas industry. Smart and intelligent fields are not just a technological upgrade—they are a necessity of how hydrocarbon assets are effectively managed, operated, and optimized to ensure that the last oil droplet and last gas molecule will be economically produced and relevant to the global economy and sustainable human development. As the oil and gas industry transition deeper into the digital era, success will hinge not only on innovation but on integration, foresight, and the ability to adapt, making the human factor a critical pillar of this evolution. The oil and gas companies that embrace this shift will be the ones best positioned to thrive in an increasingly energy-demanding, environmentally complex, and carbon conscious world.

References

1 International Energy Agency (IEA) 2025 Energy Outlook, IEA.org.

2 Nadri Pari M, Kabir AH, Mahdia Motahhari S, et al. Smart well- Benefits, types of sensors, challenges, economic consideration, and application in fractured reservoir. SPE Kingdom of Saudi Arabia Annual Technical Symposium and Exhibition, 126093, 2009.

3 Saputelli L, Duran J, Rivas F, et al. Success cases and lessons learned after 20 years of oilfield digitalization efforts. SPE Annual Technical Conference and Exhibition held in New Orleans, Louisiana, USA; 2024.

4 Adekunte PA, Sadiku MNO, Sadiku JO. Digital oil field. International Journal of Trend in Scientific Research and Development (IJTSRD) 2023;7(6).

5 Jalali M, Al-Blehed S, Al-Hajji A. Smart well technology applications for production optimization. SPE Annual Technical Conference and Exhibition; 2000; Dallas, USA. SPE-62979.

6 Al-Khelaiwi F, Birkeland R, Lee K, et al. Intelligent well technology: A review of worldwide applications. SPE Annual Technical Conference and Exhibition; 2007; Anaheim, USA. SPE-110267.

7 Several Halliburton, Schlumberger and Baker Hughes publicly published white papers on DOF concepts, deployment, case studies, and highlights mentioned in their annual review reports and websites, www.halliburton.com; www.slb.com; www.bakerhughes.com (2000–2024).

8 McKinsey “Digital in Oil & Gas” reports & Accenture, “Digital Oilfield” case studies and insights. McKinsey example: The next frontier for digital technologies in oil and gas (March 2016), Accenture & Microsoft Upstream Oil and Gas Digital Trends Survey, an annual or biennial series (e.g., 2016, 2019).

9 Several Saudi Aramco Intelligent Fields highlights published in several Saudi Aramco public articles including the company Annual Review (2017–2024), Sustainability reports and Aramco Journal of Technology, (www.Saudiaramco.com.sa).

10 Alnuaim S. DOF and reservoir monitoring, unpublished course materials and PowerPoint presentations. KFUPM; 2020–2026.

2

DOF Strategic Objectives

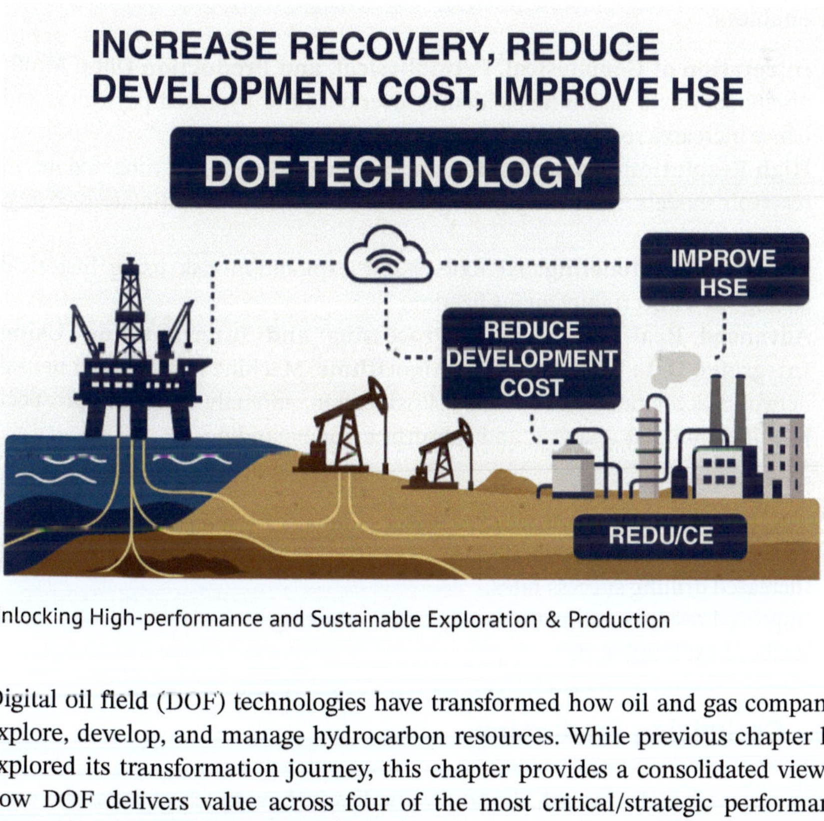

Unlocking High-performance and Sustainable Exploration & Production

Digital oil field (DOF) technologies have transformed how oil and gas companies explore, develop, and manage hydrocarbon resources. While previous chapter has explored its transformation journey, this chapter provides a consolidated view of how DOF delivers value across four of the most critical/strategic performance dimensions in upstream operations [1–4]:

1. Increasing hydrocarbon discovery and recovery
2. Optimizing production
3. Improving health, safety, environmental (HSE), and sustainability performance.
4. Reducing development costs.

Together, these corporate objectives define the operational and strategic effectiveness of any digitally enabled oil and gas exploration and production (E&P) project. The digital oil field offers an integrated platform that enables companies to pursue these strategic goals simultaneously, creating a step change in performance, resilience, and sustainability.

2.1 Increasing Hydrocarbon Discovery and Recovery

Discovery is no longer limited to seismic surveys and geological interpretation. The digital oil field provides powerful tools that accelerate prospect identification, risk analysis, and subsurface understanding.

Key enablers:

- **Integration of Geophysical, Petrophysical, and Production Data**: Multidisciplinary data fusion helps identify overlooked or bypassed pay zones and hence increases recovery.
- **High-Resolution Reservoir Models**: DOF enables near-real-time updates to reservoir models, increasing geological certainty, reserves, ultimate recovery, and production.
- **Probabilistic Modeling**: AI tools assess exploration risk using historical analogy and uncertainty quantification.
- **Advanced Real-time Seismic Processing and Interpretation Using Integrated Data Models and AI Algorithms**: Machine learning and neural networks accelerate seismic facies classification, anomaly detection, prospect identification, risk analysis, and subsurface understanding.

Impact:

- Faster evaluation of new prospects.
- Increased drilling success rates.
- Improved resource estimation and portfolio ranking.
- Reduced exploration risk.

2.2 Optimizing Production

Once a reservoir is discovered, the true value lies in how much of it can be economically and safely recovered. DOF enhances this through precision reservoir management and adaptive field development strategies.

Key enablers:

- **Real-time Production Data and Dynamic Reservoir Modeling**: DOF systems constantly update models with actual performance, enabling better decision-making.

- **Smart Completions and Intelligent Wells**: Downhole sensors and inflow control devices allow zone-specific management and optimization, resulting in increasing hydrocarbon production rates while minimizing undesired fluids such as water.
- **AI-driven Production Optimization**: ML algorithms automatically adjust electric submersible pump/gas lift settings, choke sizes, and well scheduling for maximum performance.
- **Waterflood and EOR Management Tools**: Intelligent systems control injection rates, sweep patterns, and chemical dosing for optimum operations.
- **Real-time Intelligent Reservoir Management**: Smart systems that monitor and manage reservoirs, keeping live reservoir health indices such as pressure, rate, water cut, gas–oil ratio, and sweep efficiency for maximum recovery.

Impact:

- Typical recovery factor improvement.
- Delayed onset of water/gas breakthrough.
- Increased well productivity and uptime.
- Better field-wide recovery strategies based on predictive analytics.
- Less operational cost, such as water handling, corrosion control, and compression upgrades, leading to reduced development cost ($/barrel).

2.3 Improving HSE and Sustainability Performance

In an industry where the consequences of safety and/or environmental failure can be catastrophic, DOF delivers tangible improvements in incident prediction, prevention, monitoring, and immediate response.

Key enablers:

- **Remote Operations and Surveillance**: Reduce the need for personnel in hazardous locations through automation and remote monitoring.
- **Predictive Maintenance**: Detects early signs of equipment failure before they escalate into safety incidents.
- **Leak Detection and Flare Minimization**: Real-time data identifies subsurface leaks, pipeline leaks, and excessive flaring events for immediate mitigation.
- **Emissions Detection and Minimization**: Real-time data identifies and reports emissions (methane and carbon: scope 1 and 2) and corporate sustainability indices such as carbon capture, energy efficiency, carbon intensity, and carbon circular economy.
- **AI-based Safety Monitoring**: Video analytics and sensor fusion detect unsafe behavior or environmental anomalies on site.

Impact:

- Fewer lost-time incidents (LTIs) and near-misses.
- Lower HSE exposure hours for field personnel.
- Improved compliance with environmental regulations.
- Improved compliance with corporate sustainability targets, regulations, and performance.
- Enhanced crisis readiness with early warning systems and digital incident simulations.

2.4 Reducing Development Cost

Capital and operations costs are essential, especially in volatile commodity price environments. DOF lowers costs through efficiency, optimization, automation, and real-time control across the entire E&P development lifecycle.

Key enablers:

- Faster planning cycles using integrated digital twins and collaboration platforms.
- Optimized drilling and completion through predictive analytics and automated parameter adjustment.
- Field-wide automation reducing manual labor, travel, and inefficiencies.
- Real-time operational performance dashboards identifying invisible losses and productivity bottlenecks.
- Centralized support centers (Integrated Operations Centers) enabling learner field staffing.

Impact:

- Development cost reductions.
- Operational expenditure (OPEX) savings from reduced downtime and optimized supply chains.
- Improved return on investment on marginal or complex assets.

The power of DOF lies in its ability to generate compounded value across the above 4 strategic DOF performance dimensions. For example:

- Better well placement (enabled by DOF) reduces drilling costs and enhances recovery.
- Remote monitoring lowers HSE exposure and enables faster response to field issues, improving uptime.
- Integrated asset modeling improves reservoir understanding and supports environment, sustainability, and governance (ESG) reporting through real-time emissions monitoring.

By shifting from siloed operations to a connected intelligence model, DOF turns complex trade-offs into multibenefit outcomes.

Unified value realization (Source: CERA 2010 study, Cambridge Energy Research Associates)

The oil and gas industry has considered the move toward intelligent oil and gas field operations as a technological and economic strategy. Companies that have invested in smart field technologies started reporting significant value gains in production, development cost, and HSE/sustainability performance. Based on the CERA 2010 DOF study (Cambridge Energy Research Associates), a DOF implementation typically delivers value across several areas:

- Improved recovery factor by 2–7%
- Reduced capital cost by 5–30%
- Reduce downtime by 1–2%
- Reduced OPEX cost by 5–25%

Higher production efficiency, operating cost reduction, recovery factor improvement, and emissions control have been reported in the industry. For example, the following data were obtained from various publicly published major IOC and NOC sources [5, 6]:

- Increasing production efficiency up to 10–15% through real-time optimization, smart completions, and better reservoir management.
- Decreasing in operating cost up to 20–30% by reducing manual interventions, exposure hours, and risk by automating surveillance and enabling remote operations.
- Improving ultimate recovery up to 20%+.
- Reducing capital and operating costs through drilling optimization, condition-based maintenance, and resource efficiency.
- Significantly enhancing safety through predictive and automated remote operations.
- Lowering emissions via real-time energy efficiency and methane leak monitoring.

In general, companies that master the integration of digital intelligence into every phase of exploration, production, and reservoir management will not only discover more, but they will also recover more, spend less, operate more safely, and sustainably perform better. This is the fruit of the truly and fully digitized/intelligent oil fields.

Hence, digital oil fields are a strategic enabler of performance excellence across all upstream domains, including discovery, recovery, production, safety, sustainability, and cost efficiency. As operators continue to scale their digital capabilities, DOF will play a more critical role in meeting energy demand, increasing profit, lowering carbon intensity, monitoring/reporting emissions, and safeguarding environment, assets, and people.

Moreover, as the concept of energy sustainability becomes a core driver in the energy industry, intelligent fields are becoming a major factor in supporting ESG strategic goals of corporates by improving transparency, reducing waste, and optimizing environmental/sustainability performances. This is usually monitored and managed by the CEO and Corporate Board Subcommittee.

2.4.1 Example of DOF—Saudi Aramco Haradh Oil Field

Haradh oil field of Saudi Arabia has been developed through three phases, Increment I, II, and III, each totaling 300,000 barrels of oil per day production target. Increment I was developed using vertical wells, Increment II using single horizontal wells, and Increment III using multilateral smart wells and DOF technology with inflow control valves (ICV), permanent downhole measurement system (PDHMD), and multiphase flow meters (MPFMs) [1, 5–15].

Figure 2.1 shows the development plans and their associated total development cost (X$/barrel, referenced to the vertical well development cost). The significant reduction in development cost (0.35X $/barrel) compared to the vertical well can be attributed mainly to the implementation of multilateral smart wells and DOF technology.

Multilateral smart well development cost is further reduced compared to the single horizontal brown well of almost 50% (drop from 0.7X to 0.35X $/barrel). The benefits of the intelligent field technology on Haradh field Increment III continued after the production startup (2006), maximizing the total benefits after four years of production, showing the average field water cut compared to Increments I and II, brown (none DOF) Haradh oil fields increments (Figure 2.2).

Figure 2.3 also shows the valuable benefits of the surface MPFM, subsurface ICVs, and the DOF capabilities with remote control capabilities, resulting in a significant production optimization and reduction of the well's water cut. Optimizing ICVs' downhole chokes results in a significant water production reduction. Reducing water cut will increase the well's life, increase recovery, and reduce operating cost due to less overall cost of water handling (e.g., water separation, corrosion inhibitor, special chemicals, water disposal, gas–oil separator, and water handling facility upgrades). These DOF benefits can be translated into millions of $ savings every year in both annual operating and capital expenditure, resulting in a significant reduction in development and operating costs ($/barrel).

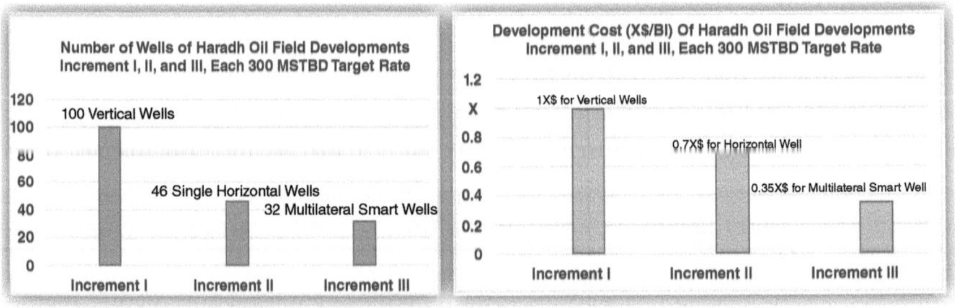

Figure 2.1 Financial benefits of Haradh oil field (plots were reconstructed from data obtained from many sources [1, 5–15]).

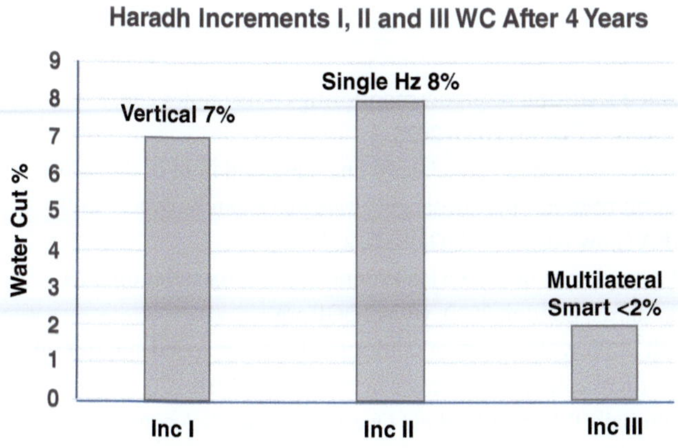

Figure 2.2 Haradh water cut (plot was reconstructed from data obtained from many sources [1, 5–15]).

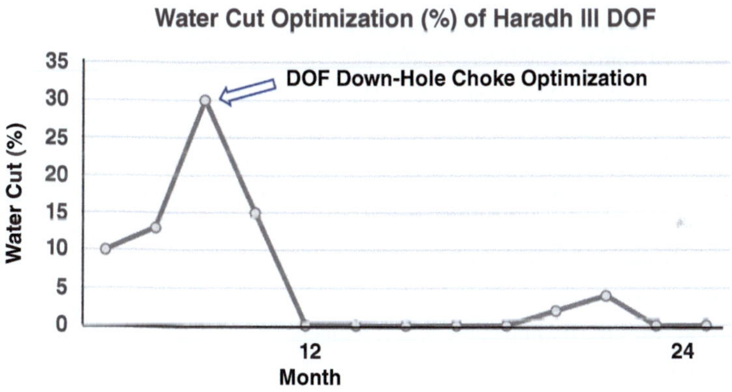

Figure 2.3 Haradh Inc III water cut optimization (plot was reconstructed from data obtained from many sources [1, 5–15]).

Smart wells and digital oil fields are no longer futuristic concepts—they are essential tools for modern petroleum engineering. Their ability to enable real-time monitoring, adaptive operations, production optimization, surface facilities optimization/debottlenecking, and data-driven real-time decision-making makes them critical to the future of energy production, especially under increasing economic and environmental pressures. Their capabilities to reduce development cost, increase production, improve safety, and corporate sustainability performances make them a necessity for any upstream oil and gas development projects. These benefits will be further explored in Chapters 7 and 8.

The following chapters will unpack the building blocks of this unprecedented, phenomenal transformation journey of the oil and gas industries, starting with the architecture and components that make digital oil fields possible.

Buckle up and enjoy the ride!

References

1 Alnuaim S. Unpublished DOF and reservoir monitoring course notes and PowerPoint presentations. KFUPM; 2020–2026.

2 CERA (Cambridge Energy Research Associates). The digital oil field of the future: enabling next generation reservoir performance; strategic research report / multi-client study, HIS CERA (now part of S&P Global), 2010.

3 IOGP (International Association of Oil & Gas Producers) reports series on data management and standards, London 2020, www.IOGP.org.

4 Accenture & Microsoft Survey. Upstream oil and gas companies spend smarter on digital technologies to drive value, reduce costs in downturn; 2016.

5 AbdulKarim A, Al-Dhubaib T, Elrafie E, et al. Overview of Saudi Aramco's intelligent field program. Paper presented at the SPE Intelligent Energy Conference and Exhibition, Utrecht, The Netherlands; 2010. Paper Number: SPE-129706-MS.

6 Several articles, white papers and highlights found in several IOCs and NOCs official annual review reports and Websites (e.g. www.saudiaramco.com; www.shell.com; www.BP.com).

7 Al-Mulhim M, Al-Kaabi A, Al-Afaleg A, et al. Haradh III – smart well field development and performance optimization. SPE Paper.

8 Al-Khaldi M, Al-Ghamdi SM, Al-Dossary MN, Al-Zawawi AS. Intelligent field implementation in Saudi Aramco: value creation and lessons learned. SPE Intelligent Energy Conference and Exhibition held in Amsterdam, The Netherlands, on February 25–27, 2008.

9 Al-Dossary FM, Al-Ghamdi AA, Al-Ahmari AS. Saudi Aramco experience with smart fields and real-time optimization. SPE Intelligent Energy Conference and Exhibition held in Amsterdam, The Netherlands, from February 25–27, 2008.

10 Bhardwaj S, Al-Dossary M, Al-Zawawi AS, Al-Ahmari AS. Real-time reservoir management in Saudi Aramco's Intelligent Fields. SPE Intelligent Energy Conference and Exhibition held in Amsterdam, The Netherlands, on February 25–27, 2008.

11 Mubarak S, Chatriwala SA, Al-Subaie FM. Lessons learned from 100 intelligent wells equipped with multiple downhole valves. Aramco Journal of Technology (JOT), Fall 2009.

12 Al-Arnaout IH, Al-Dirweesh SM, Al-Zahrani RM. Production engineering experience with the first I-field implemented in Saudi Aramco at Haradh-III: Transforming vision to reality. SPE Intelligent Energy International Conference and Exhibition; 2008. SPE-112216.

13 Abdulaziz O, Al-Kaabi AO, Nabeel I, et al. Saudi Aramco, Haradh-III: Industry's largest field development with maximum-reservoir-contact wells, smart-well completions, and the iField concept. *SPE Prod. Oper. J.* 2008;23(4):444–447.

14 Al-Mulhim A, Al-Harbi H. The role of real-time operation centres in enhancing Saudi Aramco's field performance. SPE Intelligent Energy Conference and Exhibition; 2018.

15 Al-Mulhim WA, Faddagh HA, Shehab MA, et al. SPE-128837. Mega I-field application in the world's largest oil increment development. SPE Intelligent Energy Conference and Exhibition; 2010.

3

DOF Infrastructure and Asset Management

Building the Digital Backbone of Intelligent Oil and Gas Fields

The transformation from conventional (brown) to smart and intelligent oil fields is not solely driven by algorithms, AI, or data science (as we see its astonishing harvest nowadays). It begins with a very long technology-driven foundational layer of digital infrastructure journey that the industry has started since the 1990s. Just as drilling rigs and pipelines have long served as the backbone of physical oil and gas production, the smart field is underpinned by an ecosystem of sensors, networks, computational platforms, and control systems that enable real-time monitoring, decision-making, and automation, benefiting from the digital maturity the industry has achieved in the last four to five decades.

This chapter was reconstructed from a blend of most of the references found in this domain [1–64], summarizing my understanding of the key technological components that make up a fully integrated and comprehensive infrastructure of a smart digital oil field (DOF). Furthermore, the chapter was written in a generic style, focusing more on the DOF infrastructure components and equipment functionalities without mentioning their manufacture(s) to avoid any commercial content in the book. This includes the generic typical DOF hardware, software, data, and workflow systems deployed across wells, pipelines, facilities, and corporate centers.

A well-designed DOF infrastructure is not only essential for operational efficiency but also a prerequisite for adopting advanced analytics, machine learning (ML), artificial intelligence, and the ultimate fully autonomous operations. Furthermore, the importance of the DOF assets' management and maintenance will be highlighted as a key enabler and vital component of the infrastructure, ensuring the longevity, accuracy, safety, reliability, system availability, and performance of the DOF technologies.

3.1 Intelligent Fields Architecture Overview: From Edge to Enterprise System

Smart field infrastructure is typically layered in a hierarchical architecture (Figure 3.1) composed of the following multilayers.

- **Subsurface Sensors**: This is where all subsurface data are generated and transmitted with the objective of providing real-time monitoring of every subsurface parameter(s) related to well integrity and/or well productivity, with hydraulic/electric lines control capabilities and a vision to real-time monitor and manage every droplet of oil.
- **Surface Sensors and Wellsite Device**: This is where all surface, wellhead, and wellsite data are generated, collected, monitored, and transmitted to local centers using a local network or SCADA system.

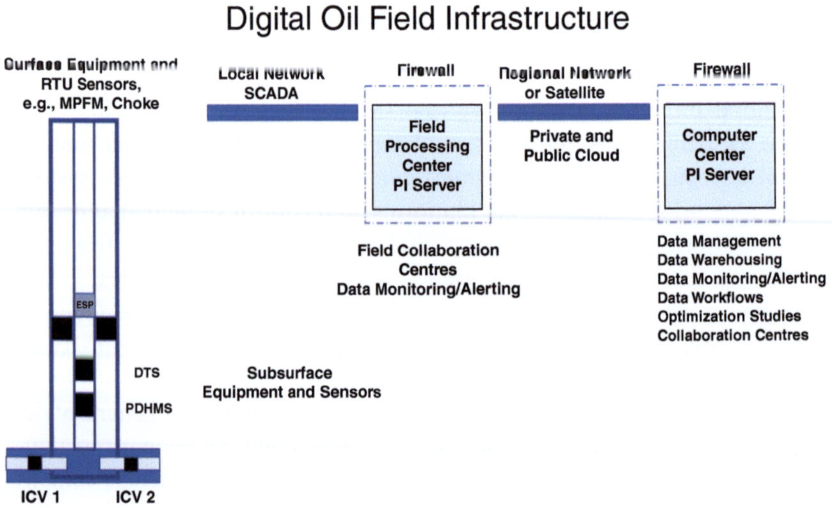

Figure 3.1 Example of a generic and typical DOF infrastructure.

- **Communication Layer**: Data collection and transmission, network and IT infrastructure (remote telemetry units [RTUs], fiber optics, radio, satellite, cellular, firewalls, private/public cloud, and Internet of Things [IoT]) that carry the generated data between field equipment and transmit it to centralized centers.
- **Data Management and Integration Layer**: Collects and standardizes data from multiple sources using middleware, APIs, and data lakes, providing quality assurance and reliable data.
- **Operational and Engineering Workflows Layer**: This layer uses the QCed real-time data and converts them into operational and engineering workflows, harvesting the value of this technology. Additional AI, ML, and advanced algorithms workflows also provide predictive capabilities to significantly increase the return on investment (ROI).
- **Control Systems and Automation**: In the context of the DOF, control and automation systems form the operational backbone that enables real-time data utilization, remote asset management, and efficient decision-making. These systems integrate sensor data, industrial distributed control systems (DCS), programmable logic controllers (PLCs), supervisory control and data acquisition (SCADA), drones, robots, and advanced analytics toward fully autonomous field operations.

Together, these layers enable the DOF to function as a connected, data-driven environment, integrated with smart business workflows and enterprise databases/systems.

3.1.1 Subsurface Sensors

At the heart of the well-based DOF capabilities are the subsurface sensors and downhole data acquisition systems, which collect critical subsurface well-specific data in real-time with the objective of providing real-time monitoring of all subsurface parameter(s) related to the well integrity and/or well productivity with control capabilities using hydraulic or electric lines. The objectives are to digitally and intelligently control the well, maximize productivity, and monitor every oil droplet from the reservoir to the wellhead. This equipment includes different types of control valves, sensors, and production devices such as electrical submersible pumps (ESPs).

3.1.1.1 Inflow Control Valves (ICVs)

Active valves that can be opened, closed, or choked with remote control capability through hydraulic or electric lines. ICVs, or sometimes called intelligent completion valves, are key components in smart well completions, used to optimize reservoir drainage from multi-reservoir zones (compartments), minimize water or gas breakthrough, and, hence, enhance oil recovery in horizontal and multilateral wells (Figure 3.2) [10–12, 17, 19, 23, 36, 59, 60].

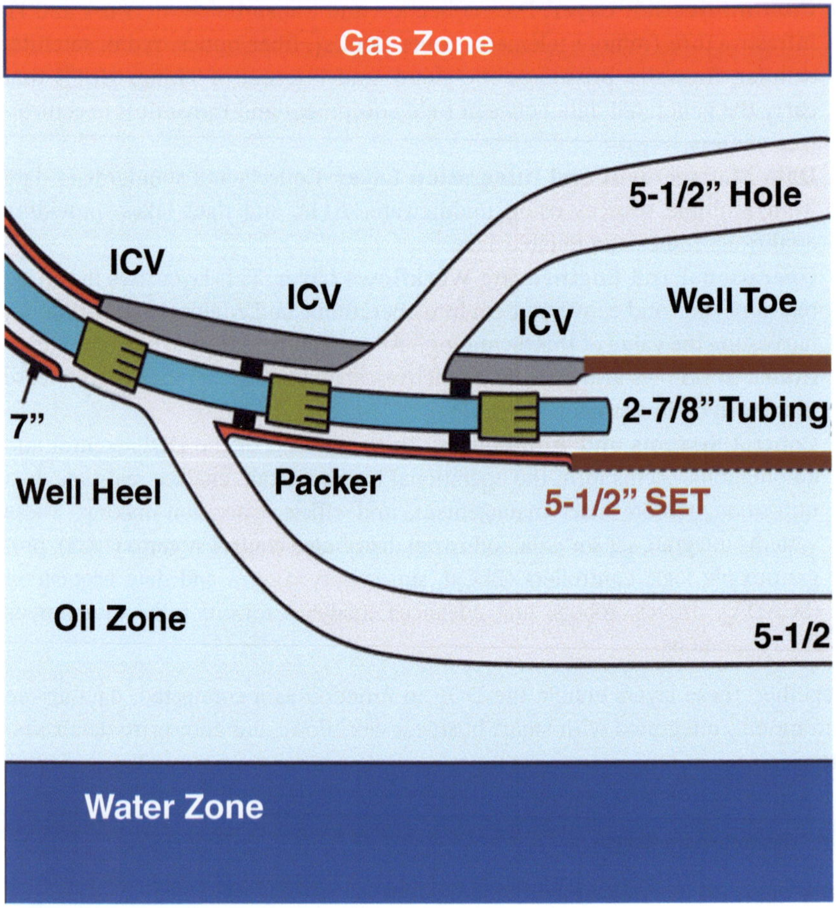

Figure 3.2 Generic ICV smart well completion in a multilateral well.

3.1.1.1.1 ICV Specifications (API, ISO, etc.)

These downhole chokes vary by manufacturer and well conditions. Downhole ICVs are simply remotely controlled downhole chokes placed in the sand face (lower completion) to control certain reservoir compartments with the objective to have better reservoir depletion and uniform sweep efficiency strategies, resulting in a longer well life, higher oil production, and ultimate recovery factor.

The choke specifications have been significantly improved since the inception of this technology almost two decades ago. Nowadays, operators have best practices to deploy, use, and maintain these advanced subsurface devices to achieve their objectives with high standards as per equipment's specification institutions such as API, IEEE, ISO, and NACE. Figure 3.3 illustrates a typical ICV configuration and different components of a typical horizontal well with two compartments and two ICV completion designs. The ICV setting is based on the well condition if it is close to the oil–water contact (OWC) and/or oil–gas contact and each

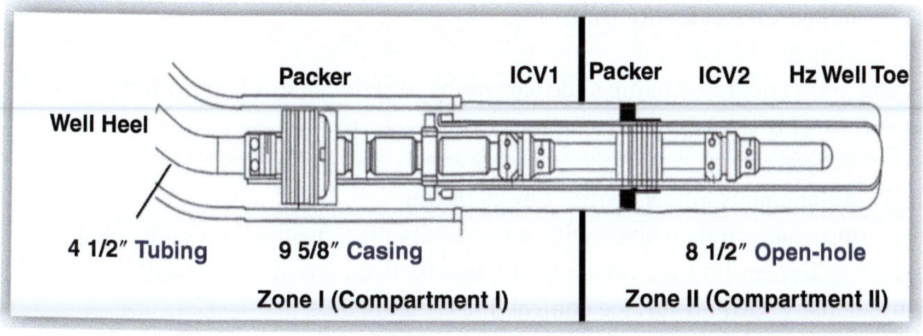

Figure 3.3 Schematic of a generic horizontal multilateral oil well with open hole smart completion with two compartments (zones) controlled by two ICVs.

reservoir-controlled-compartment productivity and permeability. The higher permeability and closer to OWC/GOC, the smaller the ICV setting (choke opening) and vice versa, ensuring better recovery and sweep efficiency.

Here's a breakdown of the types and required specifications (API, ISO, etc.) and industry standards associated with these devices:

ICVs: Active valves with surface-controlled choke openings mechanism:

- Hydraulically actuated (most common) type requiring a hydraulic line from the downhole lower completion ICV all the way up to the wellhead through subsurface packers' and surface wellheads' penetrators.
- Electrically actuated type requiring an electric line from the downhole lower completion ICV all the way up to the wellhead through subsurface packers' and surface wellheads' penetrators.

3.1.1.1.2 ICV Pressure and Temperature Ratings

A high range of pressure and temperature fits most of the world's oil field wells.

3.1.1.1.3 Bore Sizes

Typically matching all common tubing and completion sizes.

3.1.1.1.4 Control Lines

- One hydraulic or electric line per zone (lower completion-controlled compartment), which limits the number of practice ICVs used in a well to two to three max due to the limited tubing casing annulus area needed for multiple hydraulic/electric lines.
- Electric conducting technology, which can increase the number of controlled ICVs (e.g., SLB Manara™). The Manara was designed to have only one main electric line from the wellhead to the lower completion, of which power can be transmitted to all ICVs, regardless of their number, through electrical conduction technology.

3.1.1.2 Inflow Control Device (ICD)

Passive subsurface choke valves with several preset choke openings (e.g., 25%, 50%, 75%, and 100%) operated using slide sleeves by wireline operations with no control from surface (Figure 3.4).

3.1.1.3 Autonomous Inflow Control Device (AICD)

Active subsurface choke valves with a choke opening mechanism controlled by a self-floating flapper bouncy valve-closure mechanism, based on the aggregate fluid mixture density with no surface control lines (Figure 3.5).

ICV, ICD, and AICD Zonal Isolation Capability: Integrated with packer systems to control individual production zones (see Figure 3.3).

Flow Control Mechanism

- Sliding sleeve, choke, or plug-type valves (ICVs, ICD).
- Adjustable in steps or continuous positions (ICV).
- Self-floating bouncy mechanism based on the aggregate fluid mixture density (autonomous AICD).

Material Specifications

- Must be corrosion-resistant alloys such as Inconel, 13-Crome, or similar.
- Elastomers compatible with high temperature, H_2S/CO_2, and/or acidic environments.

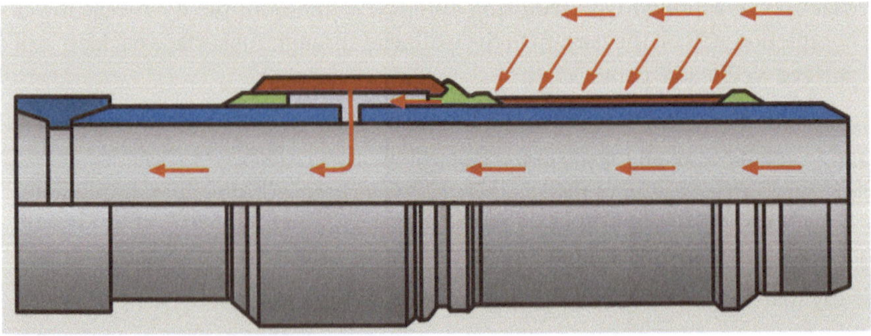

Figure 3.4 Generic AI-created inflow control device (ICD) diagram.

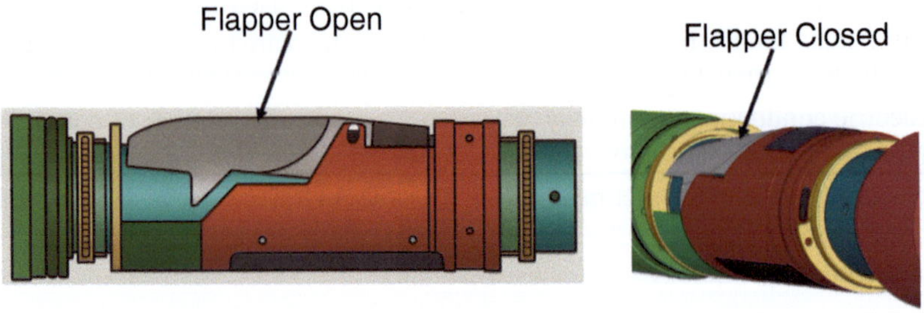

Figure 3.5 Generic AI-created diagram of an autonomous inflow control device (AICD).

Required Performance Parameters

- **Flow Control Range**: ICVs must be from 0 to 100% of open flow. ICDs may be in several preset choke openings (e.g., 25%, 50%, 75%, and 100%).
- **Zonal Control**: Full zone control with a limit of 2 to 3+ zones per well, except using a very advanced well completion system, such as the SLB Manara, providing an almost unlimited number of controlled zones (reservoir compartments or layers).
- **Response Time**: ICV must be a few seconds and the adjustable new version of ICD can be adjusted by wireline.
- **Maintenance**: The choke must be maintained by conducting periodic stroking (cycling) to ensure its operational readiness. The frequency of choke stroking varies (from a few months to annual) depending on well conditions such as sand production, corrosion, heavy hydrocarbon precipitation, scale, and CO_2 and H_2S.
- **Reliability**: Most of the industry ICVs and ICDs are very reliable, with thousands of applications deployed worldwide.

ICVs, ICDs, and AICDs are very strategic parts of the DOF, helping operators to achieve most of the DOF objectives, especially the objectives related to increasing productivity, reducing development cost, and reducing undesired fluids (water and/or gas); hence, they must be immediately commissioned, continuously maintained, and kept ready for manual or automatic zonation and well productivity optimization. Without the best maintenance practices, these advanced DOF technologies may not offer the benefits they were deployed to achieve.

3.1.1.4 Permanent Downhole Measurement System (PDHMS)

The PDHMS is a vital part of all smart and DOF strategies, providing real-time, continuous data on downhole pressure, temperature, and often other reservoir and wellbore conditions (Figure 3.6) such as density and saturation. These systems are permanently installed in the wellbore and can remain operational for the entire

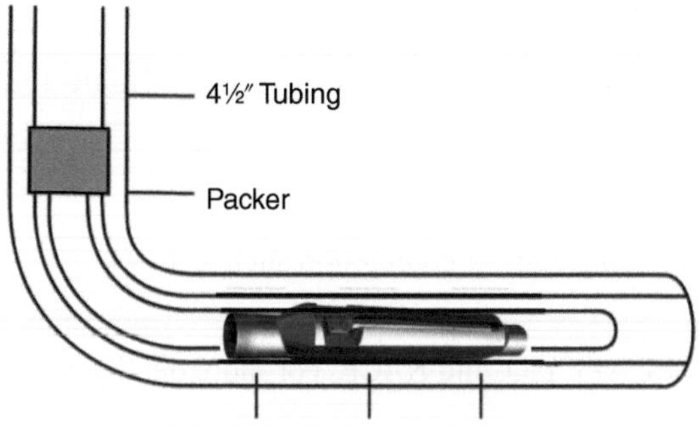

Figure 3.6 Generic permanent downhole measurement system (DHMS).

life of the well. Recent technology developments are wirelessly retrievable, which makes them replaceable [28].

PDHMS transmits pressure and temperature readings to the surface data acquisition unit (SDAU) by cable. The SDAU can capture several readings per second. The SDAU transmits this data to an RTU located near the wellhead, which connects to a central facility by satellite, wireless radio, or wired fiber optic. Then, the pressure and temperature data are transmitted to an area plant (e.g., gas–oil separator plant), where it is picked up by the area plant information (PI) system server and mapped to the proper parameter/well. This data is then quality assured by an advanced algorithm and stored in the appropriate corporate central database and applications. After this, it becomes available for use at the engineer's desktop and other operational/engineering applications and workflows.

3.1.1.4.1 Required Specifications of PDHMS for Pressure and Temperature

Measured Parameters

- **Reservoir Pressure**: All ranges of pressure seen in oil reservoirs.
- **Temperature**: Almost all ranges of temperature are seen in oil fields and possibly more for high-temperature systems.

Pressure and Temperature Sensor Types

- Quartz crystal pressure sensors.
- Resistance temperature detector (RTD) designed for a harsh downhole environment (high temperature, pressure, and corrosive fluids).
- Fiber-optic distributed temperature sensing (DTS) measuring temperature along the entire wellbore from the lower completion to the wellhead.

Power and Communication

- **Power Source**: Surface-powered via cables (common), powered through the electric grid, or preferably using solar panels or a downhole rechargeable/ replaceable battery.

Materials and Design

- **Housing**: Must be corrosion-resistant materials such as Inconel/Crome-13.

Installation Configurations

- Behind casing or tubing conveyed.
- Inside intelligent completion (part of lower completion such as ICV).
- Fiber-optic DTS.

Industry Standards for PDHMS: API, ISO, and NACE. For more details, refer to their websites.

Performance Requirements: Must be very reliable with pressure/temperature values, drift/year, and long-term (years) life expectancy.

3.1.1.5 Downhole Subsurface Safety Valves (SSSVs)

SSSVs are critical components of well completion systems in the digital oil and gas wells, particularly offshore and fields near-populated areas with high H_2S and high-risk fields (Figure 3.7). These valves are designed to automatically shut off the flow of hydrocarbons from the wellbore to the surface in the event of an emergency or wellhead failure, thus protecting people, the environment, and equipment [19, 21].

3.1.1.5.1 *Function and Importance*

SSSVs are typically installed in the tubing string of a well and are operated remotely from the surface either hydraulically, electrically, or through wirelines. Their function is part of a broader well integrity and wellsite safety strategy that includes both surface and subsurface pressure control mechanisms. It provides emergency shut-in capability just in case the wellhead integrity is at risk for any reason (isolate reservoir pressure from surface equipment). It also minimizes the risk of blowouts with automatic shutdown when a big pressure drawdown happens. It is required by industry regulations and government regulatory authorities for all offshore and near-populated onshore areas, especially fields with high pressure and H_2S.

3.1.1.5.2 *Working Principle*

Most SSSVs are normally closed valves held open hydraulically from the surface. If control pressure is lost or manually cut, a spring or gas-loaded mechanism causes the valve to completely shut down.

3.1.1.5.3 *Types of Subsurface Safety Valves*

Wireline Retrievable Safety Valve

- **Installed**: Inside a landing nipple in the tubing string (Figure 3.7).
- **Wireline Retrievable**: Replaced and/or maintained without pulling the tubing.

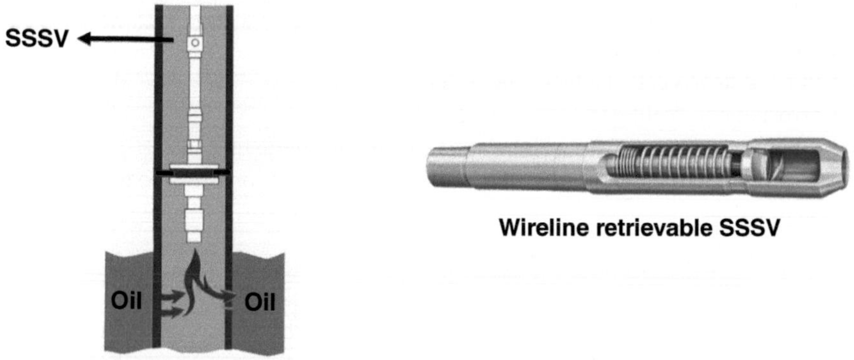

Figure 3.7 A typical schematic of a subsurface safety valve (SSSV).

Tubing Nonretrievable Safety Valve

- **Installed**: As an integral part of the tubing string.
- **Retrievable**: Only by pulling the tubing.
- **Advantage**: More reliable due to larger diameter and fewer mechanical joints.
- **Common Use**: High-reliability wells, deeper completions, supercritical wells with high rates.

Required Specifications of SSSV:

- **Size**: Matches all tubing sizes.
- **Temperature and Pressure Rating**: Must meet and far exceed all pressure and temperature ranges seen in oil fields worldwide.
- **Actuation**: Must be automatic hydraulic pressure control. The system can be integrated into SCADA for real-time monitoring and automation.
- **Flow Area**: Up to 90% of tubing ID (for minimal production restriction).
- **Response Time**: Must be a few seconds for full closure.
- **Material**: Crome-13, Inconel, or other corrosion-resistant alloys for sour (H_2S), corrosive (CO_2), and high pressure and high temperature (HPHT) systems.
- **Maintenance**: Like any other DOF equipment, SSSV must be maintained by testing and monitoring for integrity by conducting periodic manual cycling to ensure its operational readiness. The frequency varies depending on well conditions such as sand production, corrosion, and heavy hydrocarbon precipitation

Industry Standards for SSSV: API, ISO, and NACE. For more details, refer to their websites.

Emerging SSSV Technologies:

Electric Subsurface Safety Valves (eSSSVs): Use electric signals instead of hydraulic, providing faster response and smooth integration with digital well monitoring systems.

Smart Valves: Equipped with sensors and telemetry with self-monitoring and adaptive to flow/pressure anomalies.

Challenges and Failure Modes

- Control line leaks leading to valve malfunction.
- Sand and debris causing blockage or flapper wear.
- Corrosion and scaling in sour service wells.
- Mechanical failure of the spring or the sealing mechanism.
- Mitigation includes regular testing, high-quality metallurgy, and clean completion practices.

Subsurface safety valves are fundamental parts of well safety architecture. Selecting the right type, installing it correctly, and maintaining it as per best practices ensures robust well control and environmental protection, especially in remote offshore and nearby populated areas. As technology evolves, SSSVs are becoming smarter and more integrated into DOF, enabling safer and more efficient well operations' safety and environmental protection.

3.1.1.6 Electric Submersible Pump

ESPs are artificial lift systems widely used in oil wells to enhance production by lifting fluids to the surface. Installed downhole, ESPs are particularly effective in high-volume, medium-to-low viscosity wells and are a mainstay in both onshore and offshore oil production strategies [23–25].

The ESP system is electrically driven and submerged in well fluids. Due to its compact size, high efficiency, and adaptability, ESP technology is a key element in the artificial lift portfolio.

3.1.1.6.1 ESP System Overview

An ESP system typically (Figure 3.8) consists of the following main components (from bottom to top).

- **Pump**: Multistage centrifugal pump that imparts kinetic energy to lift fluids.
- **Protector (Seal Section)**: Isolates well fluids from the motor and compensates for pressure and temperature changes.
- **Motor**: Electrical motor (typically three-phase induction motor).
- **Electric Cable**: Delivers electrical power from the surface to the motor.
- **Intake/Suction**: Allows reservoir fluids to enter the pump.
- **Gas Separator**: Separates free gas before fluid enters the pump (optional in high gas oil ratio wells).
- **Working Principle**: ESPs operate using centrifugal force. The motor drives the pump shaft, which rotates the impellers. The fluid enters the pump through

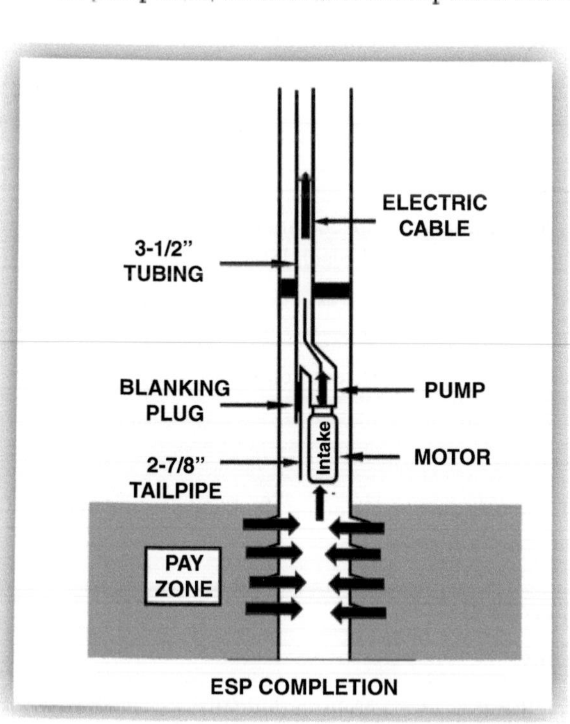

Figure 3.8 Schematic of typical Y-inverted ESP completion and design.

the intake and passes through several impeller/diffuser stages, where pressure and velocity increase. The fluid is then pushed up through the production tubing to the surface.

3.1.1.6.2 Types of ESP Systems

Standard ESP

- Basic design with motor, seal, and pump.
- Suitable for clean, moderate-depth wells.
- **Applications**: Waterfloods, horizontal wells, light oil.

High-Temperature ESP

- Designed to operate in high-temperature fields.
- Special insulation and high-temperature materials.
- Applications: Geothermal wells, high-temperature reservoirs.

Corrosion-Resistant ESP

- Materials like stainless steel and Inconel used to withstand corrosive fluids.
- **Applications**: Sour service (H_2S), CO_2-rich reservoirs.

Abrasion-Resistant ESP

- Hardened materials and special coatings for handling sand/solids.
- **Applications**: Unconsolidated sandstone formations.

Horizontal ESPs (H-ESP)

- Installed in horizontal sections of the well.
- Uses flexible cable-deployed technology or special anchors.
- Applications: Tight oil, shale plays, extended-reach laterals.

Dual ESPs

- Two ESPs installed in parallel or series for redundancy or staged lifting for high-volume or deep wells.

Cable-Deployed ESPs

- Run on power cable without using a rig for quick deployment and retrieval applications such as unconventional wells.

3.1.1.6.3 Required Design Considerations

Reservoir and Fluid Properties

- Water cut, rate, API, gas–oil ratio, productivity index (PI), gas content, viscosity, and solid concentration.

Well Deviation and Geometry

- ESPs prefer vertical or slightly deviated wells. In high-deviation wells, special centralizers or horizontal systems are used.

Electrical Load and Heat Dissipation

- Motors must be cooled by the produced fluid. Thermal derating may be needed in low-flow wells.

Gas Handling

- ESPs have limited free gas tolerance (typically <10–15% of fluid volume).

3.1.1.6.4 Monitoring and Control Systems

Modern ESP systems include digital monitoring and automation for performance and protection, including the following key monitored parameters:

- Downhole sensors for pressure, temperature, vibration, and current.
- Variable speed drives (VSDs) to adjust motor frequency and control flow.
- SCADA integration for remote control, diagnostics, and optimization.

ESP Applications in Oil Wells

- **High-volume Production Wells**: Conventional reservoirs with high permeability.
- **Heavy Oil Production**: With appropriate gas lift or thermal assistance.
- **Offshore Fields**: Compact size suits subsea completions.
- **Unconventional Wells**: As rig-less-deployable ESPs or a temporary lift solution.

3.1.1.6.5 ESP Advantages and Limitations

Advantages

- High production rates.
- Automated and remotely operated, monitored, and controllable.
- Compatible with most tubing sizes and completion types.

Limitations

- Sensitive to gas and solids.
- Requires electrical infrastructure, which could be very expensive ($$$).
- More complex and costly maintenance compared to conventional rod pumps.
- Shorter life span in harsh environments, especially if H_2S, sand production, and high temperature. The pump run life can be significantly improved through DOF ESP optimization using AI and ML algorithms, which can predict the pump failure several months before it happens with operational recommendations to prevent its failure.

3.1.1.6.6 ESP Monitoring and Optimization in DOFs

ESPs' reliability and efficiency are critical to the overall performance of a reservoir. In the context of DOF, ESPs are no longer standalone equipment—they are nodes in an intelligent system that enables real-time monitoring, failure prediction, and power-performance optimization. This section explores intelligent workflows used for ESP surveillance and highlights field-level optimization strategies to maximize production while minimizing energy consumption [22–25].

3.1.1.6.7 Smart ESP Monitoring in DOFs

Modern ESP monitoring systems integrate multiple data streams to provide real-time insights into equipment health and production behavior. Key parameters include:

- ESP intake pressure.
- ESP discharge pressure.
- Flowing wellhead pressure (FWHP).
- Pump frequency (Hz).
- ESP pump efficiency.
- Motor current and voltage.

These measurements are continuously logged via SCADA systems and analyzed for deviations, failures, or optimization opportunities.

3.1.1.6.8 Failure Monitoring and Prediction

Using real-time analytics and AI-driven models, failures such as trips, seal failures, or overload conditions can be predicted or detected early.

Components of ESP Failure Monitoring:

- Failure prediction due to several operational ad materials reasons.
- Trip detection (automatic shutoffs due to abnormal conditions).
- Motor load trend analysis.
- Temperature and vibration surveillance.
- Power consumption anomalies.

AI-based prediction models analyze historical data patterns to detect prefailure signatures and issue advisories for preventive action.

3.1.1.6.9 Optimization Approaches

Three major workflows are employed to optimize ESP performance:

Physics-based Optimization

- Uses calibrated pump curves and well models to adjust pump frequency and choke settings.
- Targets optimal pump efficiency by aligning operation with the best efficiency point (BEP).

AI-based Optimization

- ML models analyze big datasets of ESP performance to recommend operating settings that maximize efficiency and minimize energy usage.

Hybrid Approach

- Combines physical models with AI for enhanced accuracy and adaptability.

3.1.1.6.10 ESP Efficiency and PI Calculation

Real-time workflows allow production engineers to estimate:

- Pump efficiency.
- Well effective PI.

These indicators help determine whether the current ESP operation is:

- Within the optimal performance envelope as per the manufacturer's design.
- Overloaded or underutilized.
- At risk of early failure due to inefficiency.

3.1.1.6.11 Optimization Methodology

A structured AI-based advisory system is used to recommend:

- Optimal pump frequency (to improve hydraulic efficiency).
- Surface choke settings (to meet target flow while reducing load).

This advisory is computed based on:

- Target production rate.
- Reservoir and well parameters.
- Real time ESP data.
- Historical performance trends.
- Appendix D illustrates an example of an ESP AI failure case study.

Advantages of ESP Optimization Workflows

- **Increased Production**: Better drawdown, optimized reservoir contact.
- **Reduced Energy Cost**: Align operation to BEP and avoid overpumping.
- **Enhanced Equipment Run Life**: Reduced motor stress and heat generation.
- **Early Warning on Potential Failures**: Downtime prevention through predictive analytics.
- **Centralized Decision Support**: Enables field-wide optimization from control centers.

ESPs are a cornerstone of artificial lift in modern oil field development. With ongoing improvements in materials, automation, and adaptability to harsh conditions, ESPs continue to evolve as a highly efficient, high-rate lifting method.

Their versatility makes them suitable for a wide range of well conditions—from mature waterfloods to unconventional shale plays. Integrated with DOF infostructure, the real-time pump control of horsepower, speed, electric, and mechanical data provides valuable means to optimize well performance, pump efficiency, and pump-used power. Real-time ESP optimization can also be done through AI, ML, and advanced algorithms to avoid production loss, power high cost, workover cost, maintenance cost, and operating cost, resulting in a lower DOF development cost ($/barrel) due to longer pump run life and less operating cost. The future of ESP will be more self-adjusted/optimized using AI and ML powered by solar and/or renewables (smart ESP), which will significantly reduce the capital and operational cost. There is one statement I always say: the last oil droplet from the ground will be, most likely, produced through a smart ESP.

3.1.1.7 Fiber-Optic DTS

In the oil and gas industry, the integrity of wellbores and casing systems is critical for safe and efficient operations. Subsurface leaks can lead to environmental hazards, production losses, and safety risks. Traditional monitoring techniques often rely on discrete sensors or pressure/temperature tests, which can miss early-stage or low-rate leaks. The integration of DTS with distributed optical fiber technology presents a transformative solution for continuous, real-time temperature profiling across the entire wellbore (Figure 3.9), making early detection of subsurface leaks possible with high resolution and accuracy [14, 18, 52].

This subsection aims to:

- Explain how DTS systems detect and localize subsurface leaks in oil wells.
- Describe the integration of DTS with distributed optical fiber technology.
- Highlight system components, architecture, and deployment strategies.
- Provide field examples demonstrating effectiveness.
- Discuss advantages, limitations, and best practices.

3.1.1.7.1 Principle of DTS for Leak Detection

How DTS Works

- A laser pulse is sent through an optical fiber deployed in the well.
- Backscattering occurs due to temperature-induced molecular vibrations.
- Two backscatter signals—Stokes and anti-Stokes—are captured.
- The ratio of these signals provides a temperature measurement at discrete points along the Fiber.
- Using time-of-light-flight calculations, each temperature reading is mapped to a specific depth or location in the well.

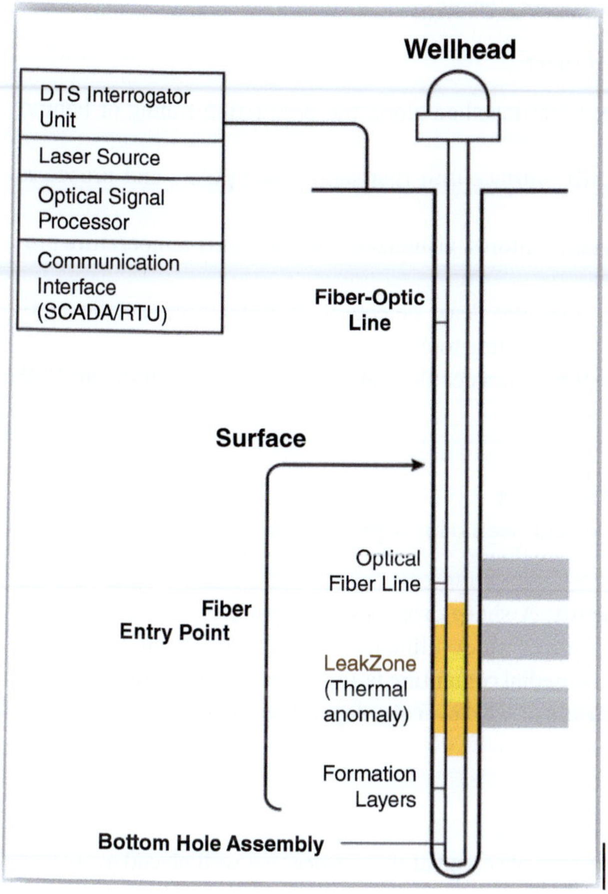

Figure 3.9 Schematic of DTS in an oil well.

3.1.1.7.2 Leak Signature in Temperature Data

Leaks create localized temperature anomalies due to:

- Fluid movement from higher to lower pressure zones.
- Heat exchange between reservoir fluids and the wellbore.
- Rapid changes in thermal gradients from gas or liquid influx.

By comparing baseline profiles with real-time DTS data, small anomalies can be interpreted as potential leak points. This is essential to prevent any undesired hydrocarbon fluids from entering shallow aquifers that could result in major environmental and health problems in addition to valuable hydrocarbon fluids loss.

3.1.1.7.3 DTS Integration with DOF in Oil Wells

Optional System Architecture:

- **Optical Fiber Cable**: Fiber installed along the production tubing or behind casing.
- **DTS Interrogator Unit**: Surface unit that sends laser pulses and processes backscattered signals.
- **Data Acquisition System**: Stores, visualizes, and interprets temperature profiles in real-time.
- **Software Interface**: Provides leak detection manufacturer-specific algorithms, event alarms, and trending tools.
- **Communication System**: Transfer the data, and integrate it into the DOF communication architecture.

3.1.1.7.4 Cases and Field Applications

Casing Leak Detection **Practical Scenario**: A production well shows signs of gas migration behind casing.

- **Example of DTS Result**: A sharp drop in temperature is detected at $\boldsymbol{x}$ feet depth, correlating with gas or water influx from a compromised annulus.
- **Example Outcome**: Remedial cementing is applied precisely at the leak zone/depth using either a rigless or workover rig, depending on the severity and risk of the special case.

3.1.1.7.5 Water Entrance in Gas Wells

Scenario: Unexpected water cut observed at the surface in a well placed away from gas–water contact.

- **Example of DTS Result**: Gradual cooling trend at $\boldsymbol{y1}$–$\boldsymbol{y2}$ feet depth indicates water entry from an upper aquifer.
- **Example Outcome**: Isolation plug set above leak; zonal integrity restored using casing patches and/or squeezed cement.

Advantages of DTS for Subsurface Leak Monitoring

- **Continuous Monitoring**: Detects both instantaneous and slow-developing leaks.
- **High Resolution**: Spatial resolutions.
- **Passive System**: No downhole power or electronics needed.
- **Real-time Alerts**: Enable proactive intervention before severe damage occurs.
- **Integration Friendly**: Seamless integration with SCADA, digital twins, and predictive analytics.

Limitations and Mitigation Strategies

- **Fiber Degradation from H_2S**: Use hydrogen-resistant or metal-sheathed Fibers.
- **Signal Attenuation at High Depth**: Use single-mode Fiber with active repeaters or dual-end access.
- **Thermal Noise**: Apply averaging or differential analysis techniques.
- **Data Overload**: Employ event-based sampling or intelligent filtering algorithms.

DTS integrated with DOF is a highly effective method for detecting, locating, and characterizing subsurface leaks in oil and gas wells. It transforms the wellbore into a fully instrumented sensor, delivering rich thermal data in real time. This capability allows operators to move from reactive to predictive maintenance, reduce environmental risks, and improve reservoir performance. As part of a broader DOF strategy, DTS will continue to shape the future of intelligent well monitoring.

3.1.2 Surface Sensors and Wellsite Device-edge Layer

Another heart of the DOF is the surface wellsite-based real-time data collection and monitoring sensors and equipment (Figure 3.11). These sensors and equipment measure wellsite/wellhead parameters such as pressure, temperature, flow rate, and choke operational parameters. This also includes real-time data for wellhead integrity, flow assurance, and other wellsite control panels for deployed subsurface technologies such as DTS fiber optics and distributed acoustic sensing (DAS), surface chemical dispensers, cathodic protection, and physical wellsite security and hence offering continuous wellbore and wellsite monitoring, integrity diagnostics, well performance, and overall wellsite security. Some DOF applications, especially in offshore and near-populated areas, include computing capability offering what is called edge computing that conducts 24/7 well integrity, H_2S sensing, emissions, and security diagnostics even during WAN interruption. These surface sensors and equipment include the following:

3.1.2.1 Multiphase Flow Meter

A cornerstone of the DOF, focusing on the integration of real-time data to enhance productivity and reduce costs, is the multiphase flow meter (MPFM). MPFM is a technology that enables the continuous, real-time measurement of oil, gas, and water produced from wells, without the need for separation (Figure 3.10) [13, 15, 26].

MPFMs are essential to optimize reservoir management, enhance production allocation, reduce environmental impact, and improve decision-making in both offshore and onshore operations. This section explores the objectives, technical specifications, operational issues, and overall benefits of MPFMs within the context of a DOF.

Figure 3.10 Typical MPFM: flow loop (generated with AI using OpenAI).

3.1.2.1.1 Objectives of Using MPFMs

The primary objectives for deploying MPFMs in a DOF environment include:

- MPFMs provide real-time flow data that allows operators to monitor well behavior and productivity continuously. This supports immediate responses to production anomalies, such as water breakthrough or slugging.
- MPFMs help in conducting production optimization, especially in multilayered and/or multilateral wells equipped with ICVs. By measuring individual well contributions without the need for test separators. Operators can optimize choke settings, artificial lift systems, and reservoir depletion strategies with the help of the real-time rate measurements provided by the MPFM.
- MPFMs eliminate the need for bulky, maintenance-heavy test separators, particularly beneficial in remote or offshore environments where space and resources are limited.
- Accurate well-by-well production data aids in fair production allocation and well rate compliance.
- Integration into DOF workflows where MPFMs feed data directly into SCADA systems, digital twins, and production dashboards, enabling AI-driven diagnostics and predictive analytics.

Types of MPFM

- **Venturi-based MPFM**: Differential pressure (DP) measurement using gamma-ray densitometry.
- **Cyclone (Centrifugal Separation) MPFM**: Partial phase separation using centrifugal force. Each phase is measured separately.

- **Coriolis-Based MPFM**: Direct mass flow and density measurement by vibration
- **Electrical Impedance/Capacitance MPFM**: Measures conductivity and permittivity of the fluid mixture.
- **Microwave/Radio Frequency MPFM**: Microwave attenuation and phase shift, measuring phase fractions directly.
- **Ultrasonic MPFM**: Transit-time velocity + phase fraction measurement.
- **Hybrid MPFM (Multi-Sensor)**: Uses venturi + gamma, microwave + DP, and/or ultrasonic + capacitance.

3.1.2.1.2 *Required Specifications and Technologies*

MPFMs are designed to operate under challenging field conditions and use a variety of technologies to measure multiphase flows (oil, water, and gas).

Required Key Specifications:

- **Ranges**: All ranges of rate, pressure, and temperature seen in global oil fields.
- **Accuracy**: Dependent on the technology and manufacturer used.
- **Flow Regimes**: Ability to measure under all common multiphase flow regimes, including stratified, slug, or annular flow conditions.

3.1.2.1.3 *Digital Capabilities*

Modern MPFMs come with digital enhancements with IoT connectivity, self-calibration, and diagnostics.

3.1.2.1.4 *Operational Issues and Challenges*

Despite their benefits, MPFMs face several challenges in practical deployment.

Calibration and Accuracy

- **Challenge**: MPFMs often require field-specific calibration due to changes in fluid composition and flow regime. Calibration is usually done periodically using either an actual test separator or a recently calibrated digital twin or well model.
- **Mitigation**: Use of smart algorithms and ML to adapt measurements over time. Self-adjusted and calibrated capability is still a dream that must be achieved in the next generation of MPFM to reduce the accuracy risk.

Flow Regime Complexity

- Some MPFMs struggle to maintain accuracy under changing or unstable flow patterns, such as slug.

Installation Constraints

- Space limitations, especially offshore, may restrict MPFM size or require custom solutions.

3.1.2.1.5 High CAPEX
MPFMs are significantly more expensive than conventional metering systems, making ROI evaluation critical, especially in brownfield projects.

3.1.2.1.6 Benefits and Impact on DOF Operations
The deployment of MPFMs brings significant long-term benefits to operators and stakeholders.

- **Enhanced Production Surveillance**: Operators can monitor well health in real time, enabling proactive interventions and reducing deferred production.
- **Improved Reservoir Management**: Accurate well-level flow data supports better reservoir modeling and helps maximize ultimate recovery.
- **Reduced OPEX**: MPFMs minimize the need for manual well testing, reduce personnel exposure, and lower equipment maintenance costs.
- **Support for Unmanned Operations**: In remote or subsea developments, MPFMs enable fully automated and unmanned well testing, aligning with the trend toward intelligent and autonomous oil fields.
- **Environmental and Regulatory Compliance**: Accurate metering ensures transparent reporting, reducing environmental risk and supporting ESG commitments.

MPFMs are not just measurement tools; they are enablers of the DOF vision. By providing accurate, real-time, and continuous multiphase flow data, MPFMs empower operators to manage reservoirs more efficiently, reduce downtime, and maximize value across the asset lifecycle. As DOFs evolve, MPFMs will continue to play a pivotal role in supporting autonomous operations, data-driven decisions, and sustainable hydrocarbon production.

3.1.2.2 Wellhead, Pressure, and Temperature Gauges
Another core of the DOF transformation is the use of surface sensors, particularly temperature and pressure gauges installed at the wellhead (Figure 3.11).

These devices are fundamental for monitoring well conditions, ensuring safe operations, and maximizing production efficiency. Surface temperature and pressure data form the basis for a wide range of critical functions, from flow modeling to equipment protection, making them indispensable tools in modern DOFs.

3.1.2.2.1 Objectives and Functional Role
The use of real-time temperature and pressure gauges at the wellhead serves multiple operational and strategic objectives.

3.1.2.2.2 Real-time Monitoring of Well Conditions
Surface gauges provide continuous insight into dynamic well behavior, allowing operators to react instantly to abnormal trends such as surges in pressure or rapid temperature drops.

Figure 3.11 Typical wellhead digital P&T gauges (generated with AI using OpenAI).

3.1.2.2.3 Flow Assurance and Safety

Monitoring surface pressure helps in the early detection of issues like hydrate formation, wax deposition, or scaling. Sudden pressure drops or spikes may indicate flowline blockage, gas kicks, or tubing leaks. Monitoring surface temperature helps in identifying the well status, whether flowing, shut-in, or dead. For example, if wellhead temperature equals the ambient temperature with specific wellhead pressure behavior, the well status can be classified as shut-in with zero rate, which helps in rate allocations in multiwell connected trunklines.

3.1.2.2.4 Basis for Production Calculations

Temperature and pressure values are required for calculating volumetric flow rates, converting surface readings to reservoir conditions, and calibrating virtual flow meters.

3.1.2.2.5 Equipment Protection

Proper monitoring ensures critical equipment like compressors, pumps, and separators operate within safe thresholds, extending equipment life and preventing unplanned shutdowns.

3.1.2.2.6 Feed to Digital Systems

Data from surface gauges feed into SCADA, digital twins, and AI systems, enabling remote decision-making, predictive maintenance, and automated optimization routines.

3.1.2.2.7 Required Technical Specifications and Sensor Types

Modern wellhead digital temperature and pressure gauges are rugged, digital-enabled instruments designed for reliability in harsh environments.

Pressure and Temperature Gauge Specifications

- **Measurement Range**: All ranges of pressures and temperatures are available with high accuracy and resolution.
- **Sensor Type**: Strain gauge, electric, and quartz crystal.
- **Material**: Stainless Steel, Inconel, or Hastelloy (for corrosive environments).

Communication and Integration

- Compatible with SCADA, DCS, and IoT platforms.
- Data may be transmitted via fiber optics, WAN, or satellite for remote sites.
- Many smart gauges support edge computing capabilities for local processing and alarm handling.

3.1.2.2.8 Integration into DOF Ecosystems

Real-time surface gauges are often integrated with other smart field systems:

- **Production Surveillance Dashboards**: Visualize real-time data trends and anomalies.
- **Alarms and Safety Systems**: Trigger automated shut-in or alerts in case of pressure breaches.
- **Virtual Flow Meters**: Use pressure/temperature to infer multiphase flow rates.
- **Digital Twins**: Support simulation of well and equipment behavior for optimization.
- **AI/ML Models**: Train predictive models for equipment failure, flow regime changes, or sand production.

3.1.2.2.9 Challenges and Considerations

While surface gauges are generally robust, certain challenges must be managed:

- **Sensor Drift and Calibration**: Over time, sensors may drift due to harsh environments. Periodic calibration is necessary.
- **Data Quality and Noise**: Temperature fluctuations from ambient conditions or sensor placement can affect readings.
- **Power Supply**: Remote installations require solar, battery, or energy harvesting systems to maintain uptime.
- **Cybersecurity Risks**: Networked sensors need protection from unauthorized access or data tampering.

3.1.2.2.10 *Future Gauges Trends–New Technologies*

The next generation of smart surface gauges is likely to include:

- Self-diagnosing sensors with predictive fault detection.
- Wireless, low-power IoT-enabled designs for easy deployment.
- Integrated AI chips for on-device data processing and decision-making.
- Smart sensor networks capable of adapting measurement frequency based on conditions.

Real-time wellhead surface temperature and pressure gauges are foundational components of the DOF. Their ability to deliver accurate, real-time insights into well performance and safety makes them indispensable in both brownfield upgrades and greenfield developments. As the oil and gas industry continues its digital transformation, these sensors will be increasingly integrated with advanced analytics, cloud computing, and automated control systems—driving a smarter, safer, and more efficient future in energy production.

3.1.2.3 Wellhead Chokes

In oil and gas production systems, the wellhead choke plays a critical role in controlling flow from the reservoir to surface facilities. Installed at or near the surface wellhead, the choke restricts flow to manage production rates, reduce pressure, and safeguard downstream equipment (Figure 3.12) [29].

Far from being just a mechanical restriction, modern chokes are integral components of automated and intelligent production systems, especially within the evolving DOF framework. This subsection explores the objectives, design and operational parameters, challenges, benefits, and future innovations in oil field choke technology.

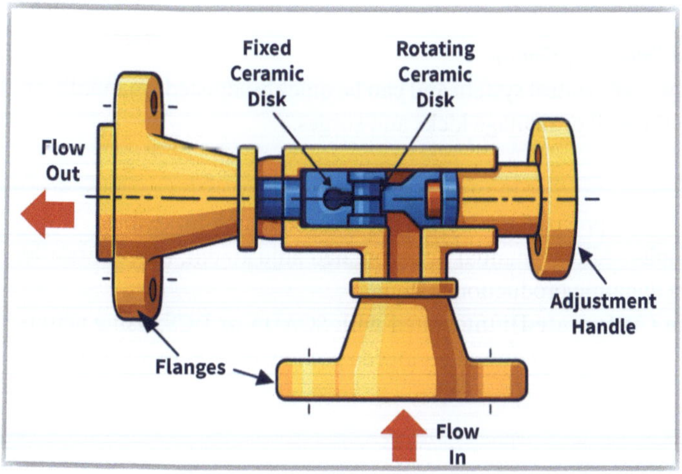

Figure 3.12 Generic schematic of ceramic-controlled choke valve (AI generated).

3.1.2.3.1 Objectives of the Wellhead Choke

The oil and gas wellhead choke serves multiple functions that directly affect well performance, surface facility integrity, and reservoir management, mainly subsurface reservoir water coning and gas cusping. This includes the following:

3.1.2.3.2 Reservoir and Flow Assurance Optimization

Controlling drawdown reduces water encroachment (breakthrough) and prolongs reservoir life, especially in mature or new depleted wells/reservoirs. In addition, restricting flow may prevent sand production and excessive drawdown, which can irreversibly damage the reservoir and negatively impact the overall oil recovery.

3.1.2.3.3 Production Optimization

In real-time operations and DOF, adjustable and smart chokes enable dynamic flow control in response to fluctuating reservoir and surface conditions.

3.1.2.3.4 Integration with Digital Systems

Modern chokes provide data-rich control, feeding into digital twins, AI systems, and automated production optimization workflows.

3.1.2.3.5 Pressure Management

Chokes help in reducing wellhead pressure, protecting surface pipelines and processing equipment from high reservoir pressures, and enabling staged pressure control in multiwell systems.

3.1.2.3.6 Production Allocation

In multiwell or commingled fields, chokes assist in equalizing flow and allocating production volumes based on contractual or reservoir management needs.

3.1.2.3.7 Safety and Emergency Control

Chokes are part of the well control system and can be quickly adjusted (manually or remotely) to shut in the well or manage kicks and surges.

Choke Types

- **Fixed Choke**: Uses a presized orifice; simple and robust but inflexible.
- **Adjustable Choke**: Allows manual or automated adjustment; more versatile and common in dynamic production systems.
- **Control Choke (Automated)**: Integrated with SCADA or DCS, using actuators for remote or automatic control deployed specially in remote and offshore fields.

Key Parameters

- **Choke Size (Orifice Diameter)**: Typically, 1/8″ or multiple of 1/8″ to 2″ or more.

- **Flow Type**: Single-phase (gas or liquid) versus multiphase (mixed oil/gas/water).
- **Operating Pressure/Temperature**: All ranges including for HPHT wells.

Operational Challenges While wellhead chokes are essential, their deployment and operation come with several technical and environmental challenges:

3.1.2.3.8 Erosion and Wear
High-velocity flow and entrained solids (sand, scale) can erode the choke trim, requiring frequent inspection and replacement.

3.1.2.3.9 Flow Instability
In multiphase flow regimes, chokes may cause or exacerbate slugging, pressure surges, or cavitation, complicating control.

3.1.2.3.10 Scale and Hydrate Formation
Cooler temperatures and pressure drop at the choke can lead to hydrate blockages or scaling, especially in gas wells.

3.1.2.3.11 Future Evolution in Choke Technology
With the rise of DOF initiatives, choke design and functionality are undergoing significant advancements.

Smart Chokes

- Digital/intelligent chokes integrate sensors and actuators with data transmission modules.
- Enable remote operation, auto-adjustment, and predictive control algorithms based on flow and pressure data.

3.1.2.3.12 AI-driven Flow Control
An AI system analyzes choke data alongside other well parameters to automatically optimize settings, improving production and reducing downtime.

3.1.2.3.13 Erosion-resistant Materials
Advanced materials like ceramic, tungsten carbide, and Inconel alloys are being used to combat wear and extend service life.

3.1.2.3.14 Wireless Control Integration
Wireless choke systems are being tested and deployed in remote or subsea wells, improving safety and reducing cabling complexity.

3.1.2.3.15 Embedded Analytics

Chokes with edge computing capabilities process data locally to trigger alerts or adjust settings without relying solely on cloud systems.

The oil and gas wellhead choke is far more than a simple valve; it is a strategic tool for managing well integrity, production efficiency, and reservoir performance. As oil fields become increasingly automated and data-driven, the humble choke is evolving into a smart, responsive component of the DOF ecosystem. Future developments in materials science, automation, and artificial intelligence promise to make chokes more resilient, more intelligent, and more integral to autonomous field operations. For operators, the key lies in understanding not just the mechanical function of the choke but its role in the broader system of safe, optimized, and intelligent hydrocarbon production.

3.1.2.4 Wellhead HSE Sensors and Equipment in DOFs

In the upstream oil and gas sector, health, safety, and environmental (HSE) protection is fundamental to sustainable operations. These sensors are critical because high-pressure hydrocarbons, environmental risks, and human activity converge at the wellsite. Hence, with the transition to DOFs, real-time HSE monitoring systems are being deployed to safeguard people, equipment, and the environment through intelligent sensing, automation, and integration (Figure 3.13) [22, 57–61].

This section explores the various required types of HSE sensors and equipment installed at the wellhead, detailing their operational roles, digital integration, and evolving future.

3.1.2.4.1 Objectives of Wellhead HSE Systems

Wellhead HSE sensors and equipment are deployed with specific operational and strategic objectives:

- Prevent hazardous events such as blowouts, gas leaks, and fires.
- Protect personnel through early warning systems and real-time monitoring.

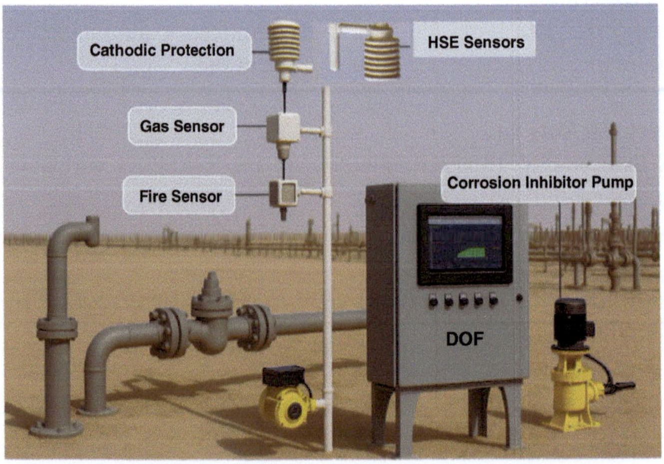

Figure 3.13 Typical wellsite with HSE sensors (generated with AI using OpenAI).

- Ensure environmental compliance by detecting emissions, spills, and unsafe hydrocarbon and greenhouse gases (GHGs) discharges.
- Enhance operational continuity by minimizing downtime from safety-related shutdowns.
- Enable automated detection, smart alarms, and real-time responses within DOF ecosystems.
- Support regulatory and reporting requirements through automatic/smart data logging and alerts.

3.1.2.4.2 Types of HSE Sensors and Equipment

Gas Detectors

- Fixed and portable detectors for methane, H_2S, CO_2, and other GHG controlled/uncontrolled emissions with real-time integration with automatic evacuation alarms and ventilation systems.

Flame and Heat Detectors

- UV/IR, multi-spectrum IR flame detectors and thermal cameras.

Pressure Relief and Emergency Shut Down Systems (ESD)

- High/low pressure sensors and actuators, valves for emergency pressure relief.

Structural Integrity Monitors

- Monitoring stress on wellhead, supports, and flowlines.

Temperature and Fire Sensors

- Thermocouples/RTDs around flammable zones and linear heat detection cables near pipelines.

Spill and Leak Detection Sensors

- Hydrocarbon leak sensors (optical, capacitive, chemical film, and acoustic emission sensors for pipeline and flange leak detection).
- Advanced applications of satellite sensing, identifying any spills/leaks, including the source and crude type, using a specific crude DNA database.

3.1.2.4.3 Applications in Oil Field Operations

HSE sensors at the wellhead are used in a range of critical applications:

- Continuous safety surveillance during drilling, completion, and production phases.
- Automatic well shut-in during gas release or fire incidents.
- Leak detection and repair programs.
- Emissions/flaring monitoring and reporting.
- Remote hazardous area monitoring using camera and drone-linked sensors.
- Worker exposure management in sour gas fields.

3.1.2.4.4 Integration in the DOF

Data Collection and Processing

- Real-time HSE data is captured via field edge devices, wireless networks, drones, satellite, and other detection devices. This data is to be integrated, in real time, with SCADA, DCS, and industrial IoT platforms.

Predictive Analytics

- ML algorithms predict safety events based on trends (e.g., gradual pressure rise, repeated gas leaks).

Alarm and Control Systems

- Sensors trigger automated shutdown sequences with geofencing and personnel tracking linked to exposure zones.

Cloud and Edge Computing

- Edge computing enables local decision-making in milliseconds and 24 × 7 service availability, even during planned/unplanned network interruptions.

3.1.2.4.5 Challenges and Limitations

Harsh Environmental Conditions High humidity, sand filling, temperature swings, and dust reduce sensor life and reliability.

3.1.2.4.6 Power Supply and Cabling

Remote installations struggle with powering multiple sensors unless solar or battery-backed systems are used.

3.1.2.4.7 False Alarms and Nuisance Trips

Poor calibration or environmental interference can result in frequent false alarms, disrupting operations.

3.1.2.4.8 Data Overload

Integrating dozens of sensors may produce too much data without proper filtering, risking missed alarms and data overload.

3.1.2.4.9 Cybersecurity Risks

Digital HSE systems, if not protected, are vulnerable to remote tampering and cyberattacks.

Benefits of Advanced HSE Monitoring

Enhanced Safety Compliance: Automated data capture ensures reporting, full traceability for audits, and incident investigation.

Reduced Operational Risk: Real-time detection enables faster response and limits escalation of incidents.

Operational Efficiency: Enables more productive operations by replacing manual checks with smart sensor-based automation.

Environmental and Sustainability Stewardship: Promotes responsible hydrocarbon production by catching leaks, spills, and gas releases early, allowing for quick response.

Workforce Protection: Toxic gas and fire detection allow for proactive evacuation and reducing employee's injury risk.

3.1.2.4.10 *Examples of Real-time Wellhead HSE Dashboards*

Here are some examples of real-time wellhead HSE dashboards and the sustainability key performance indicators (KPIs) typically monitored in DOF environments. These dashboards are usually integrated with SCADA, DCS, or cloud-based IoT platforms and allow HSE, operations, and asset integrity teams to monitor and respond to incidents in real time.

3.1.2.4.11 *Wellsite Safety Monitoring Dashboard*

Real-time visibility into personnel and gas hazard zones, including:

- Gas levels (H_2S, CH_4, CO_2).
- Wind direction and speed (for gas dispersion modeling).
- Worker location tracking (via RFID/badge).
- Evacuation zone maps.
- Live CCTV/thermal camera feeds.

3.1.2.4.12 *Wellhead Equipment Health and Alarm Dashboard*

Track the integrity and status of critical HSE equipment, including:

- Wellhead pressure and temperature alarms.
- Emergency shutdown (ESD) valve status (open/closed/failed).
- Flame detection status (triggered/standby).

3.1.2.4.13 *Environmental Compliance Dashboard*

Monitor emissions and leak events for regulatory compliance, including:

- Hydrocarbon leak detection.
- Methane emission rates (ppm or %LEL).
- Oil spill sensor status.
- Leak detection trend over time.
- Emission exceedance events and timestamps.

3.1.2.4.14 *Emergency Response Readiness Dashboard*

Ensure emergency systems are ready and responsive, including:

- Fire suppression system pressure and levels.
- Alarm system test status.
- ESD system cycle test logs.

Key HSE KPIs Monitored at the Wellhead

- **Gas Leak Frequency**: Number of gas leak detections per well/month.
- **% LEL (Lower Explosive Limit)**: Gas concentration relative to explosion risk.
- **H_2S Concentration**: Real-time toxic gas levels (ppm).
- **Wellhead Pressure Deviation**: Deviation from expected pressure ranges.
- **ESD Activation Count**: Number of emergency shutdowns (manual/auto).
- **Sensor Uptime**: Percentage of time HSE sensors are online and functional.
- **Incident Response Time**: Time from alarm trigger to system/operator response.
- **False Alarm Rate**: Percentage of false positives in alarm data.
- **Worker Exposure Events**: Number of worker toxic gas or high-noise exposure incidents.
- **Flame Detection Time**: Time from flame onset to detection and alert.

Best Practices in HSE Dashboard Design

- Color coding (e.g., green/yellow/red zones for gas levels).
- Interactive maps for sensor placement and leak/spill geolocation.
- Drill-down capability from overview to specific sensors or wells.
- Integration with alerts (SMS, email, or control room).
- Historical trend overlays for predictive analysis.
- Announced Drill exercises.

3.1.2.4.15 Future Advances in HSE Technology

AI-Driven Safety Systems: Real-time anomaly detection using AI models trained on sensor data from past incidents.

Autonomous Surveillance: Drones and robots equipped with IR, gas, and optical sensors conduct routine and emergency checks.

Satellite Remote Sensing: Satellite remote sensing technology has been used to detect leaks, emissions, and other HSE incidents with the capability to access any geographical location under any weather condition with high precision.

Digital Twin Integration: HSE data feed into digital twins to simulate safety scenarios and emergency response planning.

In an industry where safety, health, and environment are non-negotiable, HSE sensors and equipment at the wellhead are indispensable. Their evolution from analog alarms to digital-smart, connected devices forms a critical part of the DOF transformation. As oil and gas operations strive for zero-incident performance and high HSE/sustainability performance, the integration of intelligent HSE systems will continue to expand, making DOF not only smarter but also safer and more sustainable with maximum protection, especially near-populated areas and remote offshore locations.

3.1.2.5 Wellsite Device-edge Layer and Edge Computing

The DOF concept brings advanced analytics, automation, and real-time decision-making to upstream operations. At the core of this transformation is edge computing—a paradigm technology application that processes data closer to where it is generated, such as remote/offshore oil/gas well sites, rather than relying solely on network/cloud availability or remote data centers (Figure 3.14). Edge computing empowers oil and gas companies to collect, analyze, and act on data instantly, enhancing safety, reducing downtime, and improving production efficiency, especially if network coverage is provided by a third party [6, 39–42].

This section covers:

- The role of edge computing within DOF architecture.
- Key benefits of deploying Edge systems at oil well sites.
- Common hardware/software configurations.
- Real-world applications and operational use cases.
- Integration with other DOF technologies (e.g., SCADA, AI, IoT).

3.1.2.5.1 What Is Edge Computing in Oil and Gas?

Edge computing refers to processing data at or near the data source, such as wellheads, pumping stations, and separators, before it's transmitted to centralized cloud platforms.

Key Components

- **Edge Devices**: Ruggedized microservers, gateways, PCs, laptops, or PLCs installed at well sites.
- **Data Sources**: Sensors (pressure, temperature, vibration), flow meters, cameras, and so on.
- **Local Processing**: Embedded analytics, real time visualization, and anomaly detection.
- **Connectivity**: LTE, WAN, VSAT, or fiber backhaul to cloud/SCADA.

Figure 3.14 Schematic of edge computing in an oil wellsite (AI-generated image).

Benefits of Edge Computing at the Well Site

- **Real-time Response**: Low-latency detection of leaks, pressure surges, or safety risks.
- **Reduced Bandwidth Costs**: Only critical or summarized data sent to the cloud.
- **Operational Continuity**: Local decision-making even when the WAN is down.
- **Scalability**: Easily deployed across remote well pads and brownfields.
- **Security**: Limits exposure by keeping sensitive data local where possible.

3.1.2.5.2 Edge Computing in Action: Well Site Use Cases

Edge computing workflows, especially in highly hazardous and offshore locations, to monitor and assess HSE parameters with more local computing capability (edge computing) with 24/7 self/intelligent decision-making. Smart fields use edge computing to process data locally and make rapid decisions without relying on centralized systems that could be significantly affected during any network or cloud interruption. Edge computing platforms often run analytics, filtering, and AI models right at the well site or facility level. These instruments are often connected to personal computers, PLCs, and/or RTUs for local data handling and control. The following are a few examples:

1. **Pump-off Control (Artificial Lift Optimization)**
   Scenario: The rod pump needs to stop when the fluid level drops to avoid dry pumping.
   Edge Role: Real-time analysis of pump cards at the edge triggers automatic shutdown and restart without cloud latency.
2. **Leak and Pressure Anomaly Detection**
   Scenario: Rapid pressure drop detected in annular space.
   Edge Role: Edge analytics identify abnormal pressure slope, trigger emergency shut-in protocol, and notify SCADA, minimizing spillage.
3. **Environmental and Emissions Monitoring**
   Scenario: Methane sensor detects gas near surface valves.
   Edge Role: Edge gateway classifies it as a critical event and logs footage from the nearby thermal camera, forwarding flagged frames only.
4. **Equipment Health Monitoring**
   Scenario: Vibration sensors on a compressor detect unusual patterns.
   Edge Role: Edge AI compares the current signature with historical norms and flags bearing wear, enabling preventive maintenance.
5. **Severe Well Integrity Issues**
   Scenario: Well integrity issues that could end up with a blowout or wellsite fire.
   Edge Role: Local compute nodes run logic for well integrity with high severity at which the well can be fully secured by SSSV shutdown.

3.1.2.5.3 *Edge Architecture: Integration with the DOF*

Edge computing operates as part of a layered DOF stack (Figure 3.15):

Edge platforms often integrate with:

- SCADA and RTU systems.
- Digital twins for simulating well performance.
- AI/ML models that can be deployed directly on the edge.

Required Edge Hardware and Deployment Considerations

- Must operate in extreme temps, vibration, and dust.
- Preferably on solar, grid, or battery for off-grid.
- Should support intermittent sync and autonomous logic when offline.
- Use firewalls, VPNs, and device authentication for cybersecurity.
- Computers and hardware conducting edge computing at the well site can be easily robbed unless physically secured through either fences and/or remote surveillance cameras (CCTV).

Edge computing is a critical enabler of intelligent field operations in the DOF. By placing computing power directly at the oil well site, operators gain unprecedented responsiveness, autonomy, and visibility into remote operations. As production sites become more data-rich and automated, edge computing will continue to expand in scope, from controlling pumps to hosting AI models that can make smart, real-time autonomous decisions. The next edge computing generation will be focused on:

- Integrating edge devices with several strategic real-time AI anomaly detection models.
- Deploying edge-to-cloud hybrid architectures with auto-failover.
- Standardizing data formats for multi-vendor edge interoperability.

3.1.3 Communication and Instrumentation Infrastructure in DOF

The communication and instrumentation layer is the nervous system of the DOF. From field sensors to cloud analytics platforms, seamless data flow is what enables predictive maintenance, autonomous operations, and enhanced safety. At

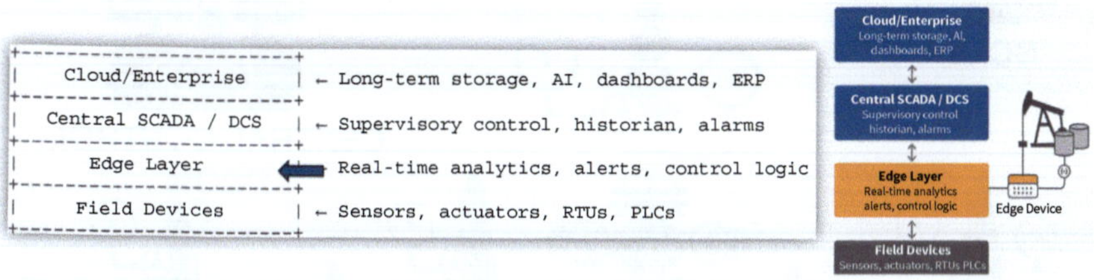

Figure 3.15 Schematic of a typical generic edge computing architecture.

the heart of DOF transformation is a robust communication and instrumentation architecture that connects hundreds of field equipment to centralized monitoring and control systems (Figure 3.16). Without a reliable communication layer and appropriate instrumentation, real-time data-driven decision-making (the essence of DOF) cannot be achieved [6–8, 53–55].

The foundation of this connectivity lies in a multitiered communication infrastructure, incorporating RTUs, SCADA systems, and satellite or wireless backhauls.

Communication systems vary based on the operator(s) specific requirements and third-party service providers. It may include:

- RTUs.
- SCADA.
- Fiber-optic networks, offering high-speed and high-volume data transmission.
- Microwave and radio links, which can be used where fiber is not feasible, such as offshore platforms.
- Cellular networks (4G/5G), enabling mobility and remote monitoring.
- Satellite communications is also essential for remote assets with no terrestrial connectivity.

This subsection details the architecture, components, communication protocols, and performance requirements for these systems to function effectively and securely.

3.1.3.1 Objectives of the Communication Layer in DOF

The primary objectives of a robust and secure communication layer in DOFs include:

- Real-time data transmission from field instruments to control centers.
- Remote control and automation of well and process equipment.

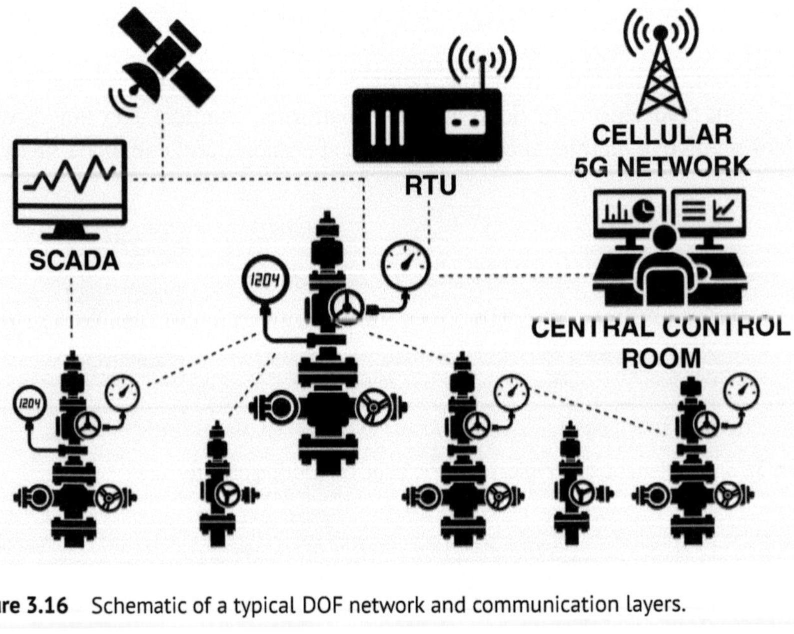

Figure 3.16 Schematic of a typical DOF network and communication layers.

- Reduced operational costs through fewer site visits and optimized manpower.
- Faster incident detection and response, improving HSE outcomes.
- Data availability for analytics, AI, and digital twins.

3.1.3.2 Core Communication Components in DOFs

3.1.3.2.1 Remote Terminal Unit

Function: RTUs collect multi sensors data and relay it to the SCADA system. They also control field devices such as valves, pumps, and ESD systems.

3.1.3.2.2 Supervisory Control and Data Acquisition

Function: SCADA systems provide centralized monitoring, control, trending, alarming, and reporting of field operations.

Architecture:

- SCADA host/server.
- Human–machine interface (HMI).
- Communication gateway.
- Historian (data archival).

3.1.3.2.3 Satellite and Wireless Backhaul

Function: Satellites and wireless networks provide long-range connectivity, especially for remote or offshore assets.

Technologies:

- **Very Small Aperture Terminal (VSAT)**: High latency but globally available.
- **4G/5G LTE**: Used where terrestrial mobile coverage exists.
- **WAN**: Low-power, long-range telemetry.
- **Microwave Point to point**: High bandwidth, low latency for site-to-site links.

Types of Communication Technologies

- **Wired Systems**: Fiber-optic cable and Ethernet.
- **Wireless Systems**: Wide area network (WAN), Wi-Fi/LTE/5G, and satellite.

3.1.3.2.4 SCADA and IoT Protocols

Network Topology Options

- **Star Topology**: Central SCADA with multiple remote RTUs.
- **Mesh Networks**: For resilience and routing flexibility.
- **Hybrid Topologies**: Combining wired and wireless for redundancy.

Cybersecurity Requirements

- Firewalls and VPNs between field and enterprise layers.
- Role based access control.
- Encryption and International compliance for industrial networks.

Reliability and Availability

- Must meet 99.9%+ uptime for mission-critical systems.
- Redundant links and dual-homed RTUs.
- UPS and backup power for field communication equipment.

Environmental Considerations

- Equipment must be rated for hazardous zones.
- Must withstand high temperature, dust, vibration, and humidity.

Challenges in Field Communication

- **Latency**: Especially in satellite links, affecting real-time control.
- **Power Availability**: Solar or battery-powered RTUs.
- **Environmental Degradation**: Corrosion, lightning, rodent damage to cabling and antennas.
- **Security Threats**: More connectivity increases attack vectors if not properly protected.

3.1.3.2.5 Future Communication Trends in DOFs

5G Adoption: Ultra-low latency supporting edge AI, video, and massive IoT.
Unified Fieldbus Protocols: Next-gen devices supporting multi-protocol stacks.
Satellite-as-a-service: On-demand bandwidth and failover support, Starlink and/or OneWeb offering high-speed satellite broadband.
Integrated OT/IT Platforms: Seamless data exchange between field automation and enterprise analytics and edge gateways, handling protocol translation and cybersecurity.

The success of any DOF depends on the strength and intelligence of its communication backbone. From RTUs collecting wellhead data, SCADA systems orchestrating operations, to satellite links enabling remote visibility, each layer must be resilient, secure, scalable, and real-time capable. As technology evolves, the communication layer is no longer just infrastructure—it's a strategic enabler of automation, safety, and efficiency across the life of an oil and gas asset. In addition to performance, network security and redundancy are critical design considerations to protect against cyber threats and ensure continuity. As oil fields continue their digital evolution, investment in resilient, intelligent, and secure networks will determine the success of long term automation, efficiency, and sustainability goals.

3.1.4 Data Management and Integration

In the DOF, data are valuable as hydrocarbons. Massive volumes of real-time information are generated from subsurface sensors, subsurface equipment, wellheads, surface equipment, pipelines, separators, and remote sensors. To transform this raw data into actionable intelligence, oil fields require a robust data management layer that organizes, validates, processes, and stores data efficiently (Figure 3.17) [3, 4, 43–48].

This subsection outlines the architecture, processes, and technologies that support effective data handling in the DOF—including collection, transmission, quality control, filtration, standardization, summarization, warehousing, and storage.

Smart fields rely on a data integration framework that can consume, clean, standardize, and contextualize information from a variety of sources:

- Real-time streaming data from sensors and control systems.
- Static data from engineering drawings, geological models, and equipment registries.
- Unstructured data such as maintenance logs, images, and video feeds.

3.1.4.1 Objectives of the Data Management Layer

The data management layer in a DOF is designed to:

- Enable real-time operational visibility.
- Ensure accuracy, completeness, and consistency of data.
- Reduce data latency and eliminate noise.
- Provide standardized data structures for enterprise-wide use.
- Supply clean, contextualized data to analytics platforms, SCADA systems, and digital twins.
- Ensure secure and scalable data storage for short-term and long-term access.

3.1.4.2 Data Collection

Sources:

- Subsurface sensors and equipment, including smart well completion valves (ICVs).
- Wellhead sensors (pressure, temperature, flow).
- MPFMs and smart chokes.
- RTUs/PLCs.
- SCADA and control systems.
- Mobile platforms (drones, wearables).
- Edge devices and gateways.

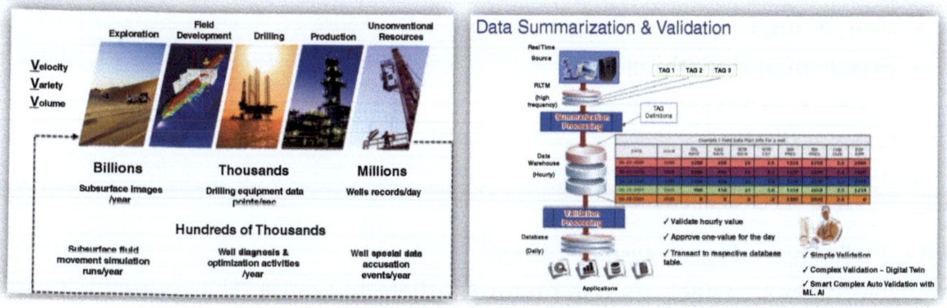

Figure 3.17 Typical DOF data management: collection, transmission, summarization, and validation (generated with AI using OpenAI).

Collection Methods:

- Polling (interval-based readings).
- Event-driven (data captured on condition change).
- Streaming (high-frequency, real-time data).

3.1.4.3 Data Transmission

Network Technologies:

- **Wired**: Ethernet, fiber optics, and so on.
- **Wireless**: Wi-Fi, 4G/5G, WAN.
- **Satellite**: VSAT or low earth orbit (LEO) for remote locations.

Transmission Requirements:

- Low latency (for control systems).
- Redundancy (dual paths, backup links).
- Security (TLS encryption, VPN, firewalls).
- Time-synchronization.

Data must be securely and reliably transmitted from the field to central platforms, cloud systems, or edge aggregators.

3.1.4.4 Data Filtration

Filtration removes unnecessary or noisy data to improve system performance and relevance.

Filtration Techniques: (Al-Gharbi 2022 [47])

- **Threshold-based Filters**: Eliminate data within deadbands.
- **Data Thinning**: Reduce data resolution from seconds to minutes.
- **Event Filtering**: Transmit only when data changes significantly.
- **Compression**: Use of delta encoding or binary formats to reduce size.
- **Other Advanced Filtration Techniques**:
  - Kalman filtering
  - Moving average
  - Kernel regression
  - Median filter
  - Exponential smoothing
  - Lowess (Savitzky Golay)
  - Wavelet filtering
  - Polynomial regression

3.1.4.4.1 Example–Moving Average Method

Moving averages are widely used to remove outliers and noise. The approach is also able to smooth fluctuations. It is based on specifying a window [1 to i], where i must be less than the size of the dataset, and the average for the window is assigned to the data point 1. After that, the window is moved to [2 to $(i + 1)$], and the average of that window is assigned to the data point 2, and this continues until it reaches the size of the dataset. The algorithm (Figure 3.18) presents the moving average algorithm's pseudocode.

3.1.4.5 Data Standardization

Data from various sources must be harmonized into consistent formats and semantics for integration.

Why It Matters:

- Enables data sharing across departments, systems, sensors, and applications from multi-vendors.
- Supports AI/ML applications with clean, labeled inputs.
- Enhances regulatory reporting and compliance.

Standardization Processes (Al-Gharbi 2022 [47]):

- Unit normalization (e.g., psi vs. bar for pressure).
- Tag mapping to enterprise naming standards for every measured and transmitted parameter.
- Data Typing: Converting analog/digital values to standardized schemas.
- Meta-tagging: Adding location, time zone, and asset context.

3.1.4.6 Data Summarization

Summarization transforms high-frequency data into readable, actionable insights.

Types of Summarizations (Al-Gharbi 2022 [47]):

- **Statistical**: Mean, max, min, standard deviation over intervals.
- **Temporal**: Hourly/daily/monthly production or emissions summaries.
- **Event-based**: Logging equipment runtimes, alarms, safety triggers.
- **Trend Extraction**: Rolling averages, rate of change, cumulative production.

Summarized data are used for dashboards, KPIs, performance reviews, and executive reporting.

Input: *Raw Real-Time Drilling Dataset*
Output: *Cleansed Drilling Real-Time Dataset*
P $\in$ *Dataset Points*
p' $\leftarrow$ *Empty Dataset*
Window Size $\leftarrow$ *Specific Moving Average Window Size*
For *i* $\in$ *P* **do**

$$P'_i \leftarrow \frac{1}{\text{Window Size}} * \Sigma^{i}_{(i+Window\ Size)}\ P_i$$

End
Return *P'*

Figure 3.18 Example of moving average filtration method (with adapted from [47]).

3.1.4.6.1 Example A: Data Filtration, Conditioning, and Summarization (Al-Gharbi 2022)

Figure 3.19 shows three different stages of real-time data collected from DOF. Plot #1 shows high-frequency, real-time transmitted gas rate collected during 24 hours from MFPM with one-second frequency, totaling 86,400 gas rate values with spikes up/down. These raw data practically cannot be used as is for any engineering or operational applications and must be quality assured, filtered, conditioned, and summarized. Plot #2 shows filtering, conditioning, and cleansing the data, including out-of-range spikes. This also cannot be used for any engineering or operational uses. Plot #3 shows the final hourly conversion of the 86,400 data points into a total of 24-hourly gas rates that can be stored and used in all engineering and operational applications and workflows. This data can also be further converted into daily average rate values for more advanced applications, such as reservoir simulation. Each stage has its own algorithm based on its engineering, operational, and corporate-specific correlations and algorithms. This can also be further improved using very advanced AI and ML algorithms, as can be seen in the next example [48].

3.1.4.6.2 Example B: Oil and Gas Summarization

Figure 3.20 shows a typical hourly oil and gas rate (A), converted to daily average oil and gas rates (B), monthly average oil and gas rates (D), and quarterly average oil and gas rates (C) using very advanced statistical algorithms.

There are several applications and workflows, such as long-term reservoir simulation and field development, that require quarterly and/or monthly rates due to their specific formulation, which may experience numerical instability if high frequency (secondly, hourly, or daily rates) are used. Other applications, such as decline curve analysis, economic feasibility studies, and other HSE applications, require daily average rates. Well testing applications will require high frequency (secondly rate and pressure data).

3.1.4.7 Data Quality and Control (QC)

Raw, filtered, conditioned, and summarized data from the field must be validated before it is trusted for decision-making.

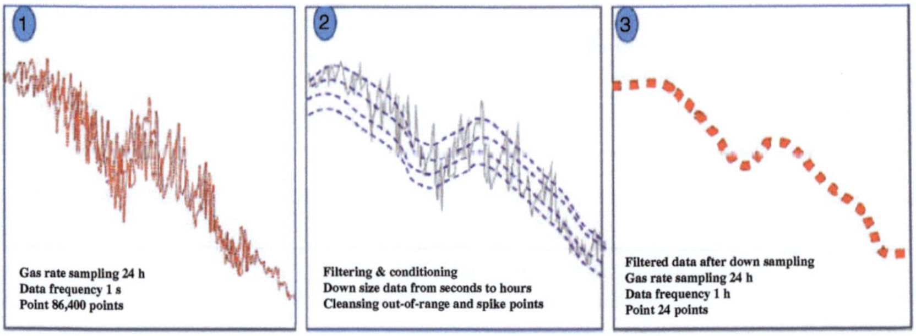

Figure 3.19 High-frequency gas rate in 24 hours with spikes and out-of-range values (1); filtered, conditioned (2), and summarized (3). Adapted from [47].

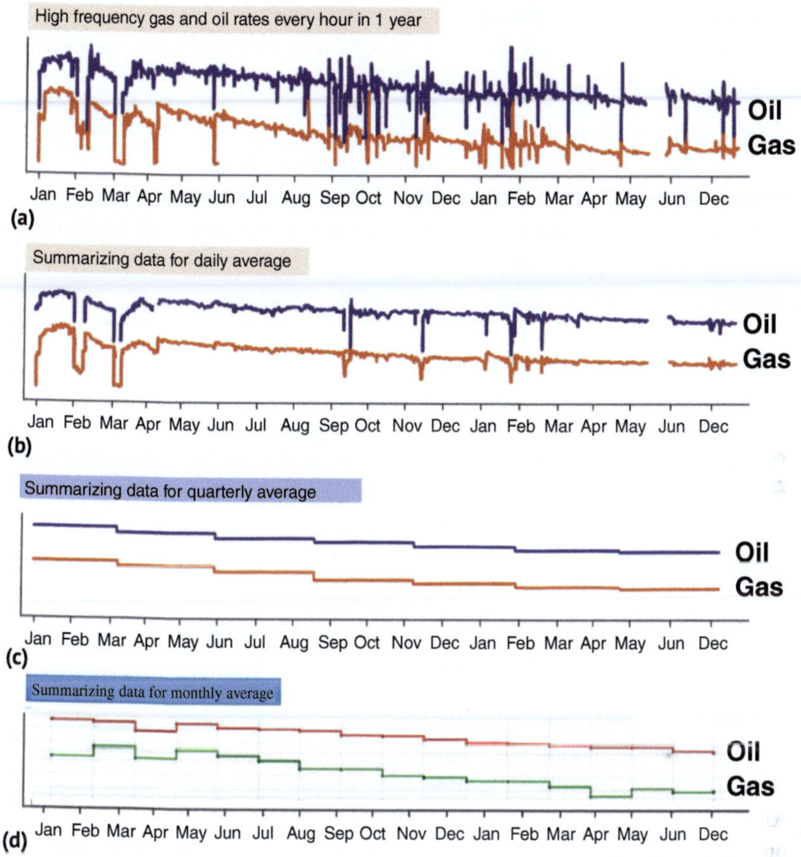

Figure 3.20 Typical hourly (a), daily (b), quarterly (c), and monthly (d) oil and gas averaging rate. Adapted from [47].

QC Processes:

- **Validation**: Range checking, missing value detection.
- **Error Correction**: Smoothing, interpolation.
- **Redundancy Checks**: Cross-verification from multiple sensors and/or virtual models.
- **Time-series Alignment**: Ensuring chronological accuracy.
- **Anomaly Detection**: Flagging sensor drift or outliers.

QC Tools:

- Edge analytics engines.
- Historian validation modules.
- ML algorithms for pattern recognition.
- Digital Twin and recently validated and updated well models.

3.1.4.8 Data Warehousing and Storage

Once validated and standardized, data are stored for operational uses, regulatory compliance, and long-term analysis (table including the following):

- Real-time cache database for fast access for SCADA/HMI/HSE applications.
- Time-series historian for operational archives and trending.
- Relational databases for structured, queryable records and information.
- Data lakes for raw, semistructured big data storage, providing efficient, quick data access to DOF applications.
- Cloud storage for scalable, long-term, accessible anywhere.

3.1.4.9 DOF Data Management Key Requirements

- **Scalability**: Petabyte-level storage potential.
- **Redundancy**: Backup and replication across regions.
- **Retention Policy (Data Life Cycle)**: Based on regulatory and operational requirements.
- **Access Management**: Secure, role-based access to sensitive data.
- **Cybersecurity**: Firewalls, DMZ, and Encryption.

3.1.4.10 DOF Data Management Integration with Other Digital Systems

Validated and stored data is made available to:

- AI/ML platforms for predictive analytics.
- Digital twins for simulation and modeling.
- Asset performance systems for maintenance insights.
- SCADA and HMI systems for live operational views.
- Compliance engines for emissions, environmental, safety, and audit reporting.

3.1.4.11 DOF Data Access and Visualization

The successful collection and quality control of measurement data stream leads to value-added datasets. The next challenge is the graphical display of sensor data, giving the user the ability to surf large datasets.

Example: The visualization systems of the DOF real-time and summarized data provide production and reservoir engineers with the functionality to monitor, diagnose, and make decisions about their wells, reservoir, or facility performance. Figure 3.21 is an example of a typical DOF well model dashboard integrated with validated real-time data, providing engineers with different views combining rates and measured or calculated pressures response for oil producers, water injectors, and water supply systems at various levels of detail.

Visualization systems can monitor the overall well/reservoir/field performance. This view can help engineers to identify quickly any deviation from strategy rate guidelines by highlighting in different colors wells that are being operated beyond their recommended strategy. Moreover, engineers can visualize the current real-time well surface and subsurface performance.

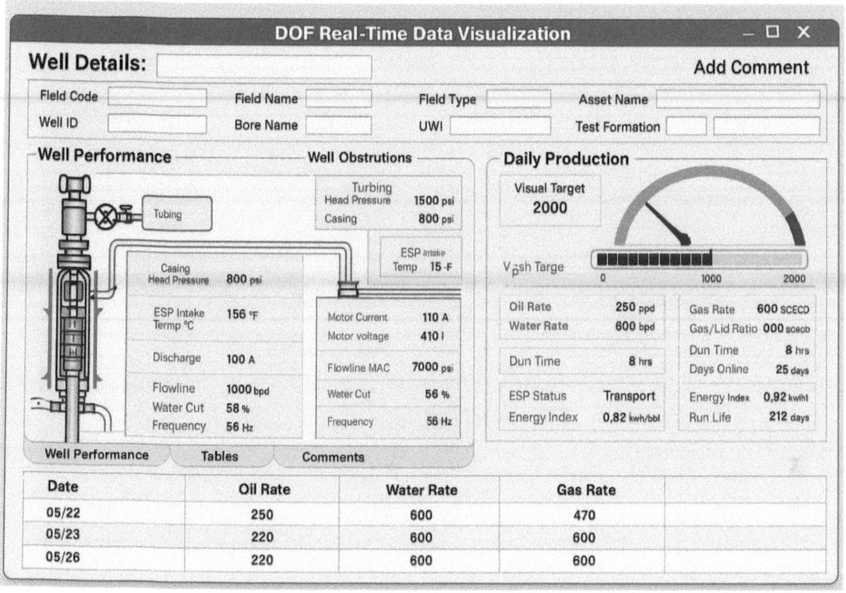

Figure 3.21 Typical AI-generated DOF real-time well-based visualization dashboard.

Data Access is also a very important part of the DOF Data Management System, where only authorized staff can visualize the data they need to see based on their organizational role and responsibility. The data owner is usually the manager of the DOF Data Access Module.

3.1.4.12 DOF Data Management Dashboards

DOF data quality from sensors all the way to the desktop can be monitored in real time and reported through live dashboards (Figure 3.22).

This live well/field-based dashboard has different statuses (based-colors) and drill-in capability to identify the data quality issues in terms of type and location. These dashboards are managed by the data owners and DOF engineers with the objective of reaching near 100% desktop data connectivity, quality, availability, and accuracy. Green color usually indicates healthy, orange medium, and red bad data quality. DOF engineers can focus more on red-colored wells/fields/equipment/data.

3.1.4.13 Challenges in DOF Data Management

- **Sensor Calibration Drift**: Invalid data infiltrates storage.
- **Data Overload**: High storage and processing costs.
- **Network Bottlenecks**: Incomplete or delayed data transmission.
- **Integration Complexity**: Legacy versus modern data format conflicts.
- **Data Security and Access Control**: Risk of unauthorized access or tampering.
- **Cybersecurity**: Risk of unauthorized access or tampering.

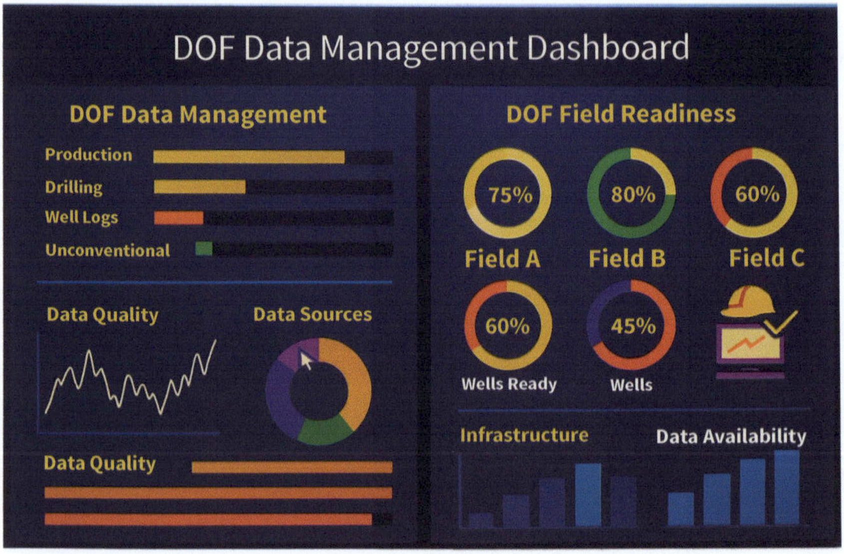

Figure 3.22 Typical live DOF overall reliability data management dashboard.

3.1.4.14 Future Trends in DOF Data Management

- Edge-to-cloud orchestration with AI-powered QC and analytics.
- Unified semantic models for seamless integration.
- Blockchain for data immutability and traceability.
- Self-service data portals for engineers and analysts.
- Autonomous data pipelines that adapt and evolve in real time.

The data management layer is the engine room of the DOF. It automatically (with no manual intervention) and smoothly transforms raw field inputs into reliable, standardized, and structured information that fuels every layer of modern operations and smart DOF applications, from real-time control to predictive analytics. A strong foundation in data collection, transmission, QC, filtration, standardization, summarization, and warehousing ensures that digital oil and gas field operators can make smarter, safer, and faster decisions. Proper data architecture is the foundation for advanced analytics, ML, and predictive models, hence helping oil and gas companies to achieve the DOF three strategic objectives of increasing production/recovery, decreasing development cost, and enhancing HSE/sustainability performance. The following are good examples of predictive algorithms that can be done automatically using proper DOF data:

- ESP failure prediction.
- Predictive maintenance such as MPFM calibration.
- Predictive subsurface and surface leaks detections.
- Predictive well safety and integrity detections.
- Predictive well performance capabilities.

3.1.5 Control Systems and Automation

In the context of the DOF, control and automation systems form the operational backbone that enables real-time data utilization, remote asset management, and efficient decision-making. These systems integrate sensor data, industrial distributed control systems, PLCs, supervisory control, data acquisition (SCADA), drones, robots, and advanced analytics to enhance oil and gas production, reduce downtime, and ensure safety (Figure 3.23) [33, 36, 54, 64].

This subsection explores the core components, architecture, and role of control and automation in the DOF infrastructure, detailing how it contributes to operational excellence and digital transformation in upstream operations.

3.1.5.1 Role of Control and Automation in DOF

Control and automation in the DOF enable the monitoring, regulation, and optimization of processes throughout the upstream value chain, from drilling and well completion to production and artificial lift.

Key benefits include:

- **Remote Operations**: Reduced need for on-site personnel via centralized command centers.
- **Safety and Compliance**: Real-time alerts and automatic shutdown protocols to prevent accidents.
- **Operational Efficiency**: Reduced latency between data acquisition and control action.
- **Predictive Control**: Integration with AI/ML models to forecast equipment behavior and prevent failures.

Figure 3.23 Control systems and automation of DOF (AI generated).

3.1.5.2 Core Components (Figure 3.24)

- **Field Instrumentation**
  The foundation of automation lies in smart sensors and transmitters scattered across the entire field that continuously monitor variables such as pressure, temperature, flow rate, HSE parameters, and vibration. These devices are often integrated with digital communication protocols like HART, Modbus, or Foundation Fieldbus and managed by operational control rooms located within the vicinity of the field, such as gas–oil separator plants), gas manifolds, gas plants, water injection plants, pipelines, and stabilization plants.
- **Local Control Systems**
  PLCs and RTUs provide localized control and data preprocessing. These are designed for harsh environments and are often deployed at wellheads, separators, and pumping stations.
- **Supervisory Control and Data Acquisition**
  SCADA systems serve as the supervisory layer that aggregates data from multiple field devices and provides an HMI for real-time visualization and control. SCADA systems are essential for:
  - Alarm and event management.
  - Data trending and historical archiving.
  - Integration with enterprise systems (e.g., ERP, maintenance management).
- **Distributed Control Systems (DCS)**
  For centralized control of complex production facilities (e.g., offshore platforms or large central processing facilities), DCS platforms provide tighter integration, more deterministic performance, and advanced process control (APC) capabilities.

3.1.5.3 Integration with Digital Platforms

Modern DOF architecture integrates control systems with digital platforms via edge computing and industrial IoT frameworks. This enables:

- **Real-time Analytics**: Edge devices process data locally and send filtered insights to the cloud.

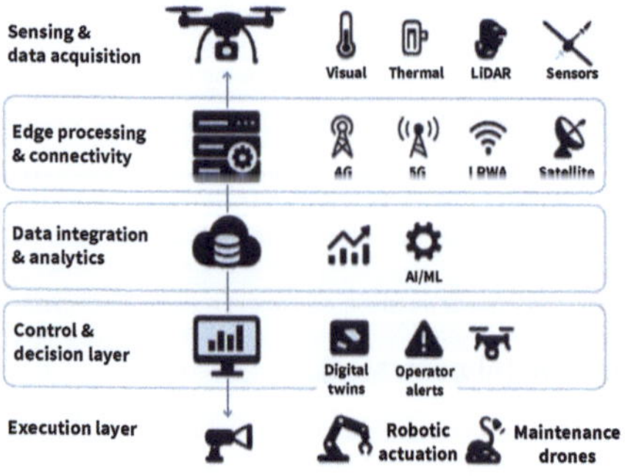

Figure 3.24 Components of DOF control and automation (AI generated).

- **Digital Twins**: Virtual representations of physical assets are synchronized with live data for predictive modeling.
- **Closed-loop Automation**: Data from cloud-based analytics can feed back into control systems to autonomously adjust operations.

3.1.5.4 Cybersecurity Considerations

As more critical field operations are digitized, cybersecurity becomes a frontline concern. Security must be embedded at every layer of infrastructure—from sensor to cloud [9, 38]. Best practices include the following:

- Segmentation between IT and operational technology (OT) networks to minimize attack surfaces.
- Firewalls, intrusion detection, and endpoint protection against unauthorized access.
- Intrusion detection systems tailored to industrial protocols.
- Patch management and vendor access control.
- Regular audits and compliance frameworks aligned with standards like NIST, ISA, and ISO.
- Incident response planning to detect, contain, and recover from breaches.

3.1.5.5 Challenges and Mitigation Strategies

Several challenges can be addressed, including the following:

- Legacy system integration, addressed using protocol converters and gateways.
- Data latency in remote areas, which can be addressed using Edge computing and satellite backhaul.
- Skill gaps in automation technology, which can be addressed through upskilling and training programs.
- Vendor lock in, which can be addressed through the use of open standards and interoperability frameworks.

3.1.5.6 The Role of Drones and Robots in DOF

The upstream oil and gas sector is undergoing a significant transformation under the umbrella of DOF practices, driven by advances in the IoT, data analytics, automation, and robotics. In this context, unmanned aerial vehicles (UAVs or "drones") and robotic systems (including ground and subsea robots) are emerging as critical enablers of inspection, monitoring, maintenance, and environmental safety functions. This subsection presents a comprehensive review of the deployment of drones and robots in oil and gas field operations, proposes a conceptual architecture for their integration, discusses benefits and challenges, and presents future directions. Key findings include that drones can reduce inspection times by up to 90% and maintenance costs by up to 50% (Alkhaldi, B. W., Alzubaidi), robots can operate in hazardous or remote environments previously inaccessible to humans, and combined systems allow real-time data acquisition supporting predictive maintenance and decision-making (Figure 3.25) [63].

Figure 3.25 Schematic of drones and robots in DOF (AI generated).

Major challenges include data integration, cybersecurity, regulatory issues, and ensuring operational reliability in harsh environments. The subsection concludes by recommending strategic adoption pathways and further research into AI-driven autonomy and swarm robotics in oil and gas fields.

3.1.5.6.1 Drones: Eyes in the Sky

Drones, or UAVs, provide a unique aerial vantage point that traditional inspection and monitoring techniques cannot match.

Inspection and Surveillance

- **Routine Site Inspections**: Drones can cover vast areas quickly, capturing high-resolution imagery and thermal data to detect equipment wear, pipeline leaks, or structural anomalies.
- **Emergency Response**: In the event of gas leaks, fires, or other hazards, drones can provide real-time situational awareness without risking human life.

Environmental Monitoring

- **Methane Leak Detection**: Specialized sensors allow drones to detect GHG emissions, aiding compliance with environmental regulations.
- **Land and Wildlife Monitoring**: UAVs help evaluate the ecological impact of drilling operations.

Logistics and Inventory

- **Remote Delivery**: In large, remote fields, drones can be used to transport lightweight tools or materials.
- **Inventory Management**: Aerial scans of storage yards and tanks streamline inventory tracking.

3.1.5.7 Robots: The Mechanical Workforce

Robotics in DOF refers to autonomous or semi-autonomous ground units designed to perform tasks ranging from repetitive labor to high-precision operations.

Equipment Maintenance

- **Valve and Pipe Maintenance**: Robots equipped with manipulation arms can perform tightening, loosening, or lubrication tasks on wellheads and pipelines.
- **Drill Floor Operations**: Robotic systems can handle drill pipe alignment, connection, and disconnection—tasks that are both dangerous and physically demanding.

Subsurface and Hazardous Work

- **Confined Space Inspection**: Robots can enter tanks, risers, or deep shafts to assess integrity using cameras, ultrasonic sensors, or LIDAR.
- **Underwater Operations (ROVs)**: For offshore platforms, remotely operated vehicles perform seabed surveys, wellhead inspections, and repairs.

Predictive Maintenance and AI Integration

- Robots paired with AI can identify subtle signs of failure in equipment—vibrations, temperature shifts, or corrosion—enabling proactive maintenance.

3.1.5.8 Advantages of Automation in DOF

Advantages of automation in DOF include the following:

- **Safety**: Minimizing human exposure to high-pressure zones, toxic gases, and explosive environments.
- **Efficiency**: Drones and robots operate continuously, unaffected by fatigue or environmental extremes.
- **Data Quality**: Automation ensures consistent and high-frequency data collection for more accurate decision-making.
- **Cost Reduction**: Reduced labor costs, fewer accidents, and lower downtime yield significant savings.

3.1.5.9 Industry Case Study

Saudi Aramco (Aramco Elements Magazine, 2024) has used over >100 drones across facilities, including vertical takeoff drones and ultrasonic drones for non-destructive tests (NDTs), resulting in a remarkable inspection time cut by 90% and carbon intensity reduction by 31.8% between 2019 and 2022. Other major national and international oil and gas companies reported similar successes.

3.1.5.10 Challenges and Considerations

- **Regulatory Hurdles**: Airspace control, data privacy, and safety protocols must be addressed.
- **Initial Investment**: High upfront costs and the need for training personnel can be a barrier.
- Standardization of data collected by both drones and robots.
- **Interoperability**: Seamless integration with legacy systems remains a technical challenge.

- As AI and ML evolve, the synergy between intelligent drones and adaptive robots will redefine how oil field operations are conducted. Companies that invest now in robotics and drone infrastructure are positioning themselves at the forefront of operational excellence.

Drones and robots are not merely auxiliary tools in the DOF landscape. They are catalysts for a safer, smarter, and more sustainable future. Their integration into everyday operations is not a question of "if" but "when." Companies that lead this transformation will shape the next chapter of the energy sector and distributed optical fiber technology.

3.1.5.11 Future Trends of Control Systems and Automation

- **Autonomous Operations**: Using AI to enable fully autonomous drilling and production.
- **5G in Remote Control**: Enabling ultra-low-latency control in remote locations.
- **AI-enhanced Process Control**: APC driven by ML algorithms.
- The integration of drones and robots into the DOF infrastructure.
- The integration of real-time satellite imaging/sensing/monitoring into the DOF infrastructure.
- **Standardized Platforms**: Migration toward open, interoperable control system architectures (open architecture).

Control and automation systems are a critical enabler in the DOF ecosystem. Their integration into DOF infrastructure not only increases operational efficiency and safety but also lays the groundwork for next-generation innovations like autonomous operations and cognitive analytics. A well-architected automation layer transforms raw data into actionable intelligence—delivering on the promise of the DOF.

3.1.5.12 Scalability and Standardization

One of the challenges in deploying smart field infrastructure is achieving scalability across a diverse asset portfolio. Companies must develop infrastructure that is:

- **Modular**: Easy to deploy, expand, or replace.
- **Vendor-agnostic**: Compatible with equipment and software from multiple providers.
- **Standardized**: Uses common data models, communication protocols, and interface designs.
- **Cost-effective**: Balances technical capabilities with ROI across new and existing assets.

Standardization reduces deployment times, training costs, and integration risks.

3.1.5.13 Sustainability and Energy Efficiency

Smart field infrastructure also contributes to environmental performance:

- Energy-efficient equipment (e.g., variable speed drives, efficient compressors).
- Emission sensors to detect and minimize flaring and fugitive gas releases.

- Real-time environmental monitoring for water usage, spills, and ecological impact.
- Power management systems to optimize energy consumption at remote facilities.

As ESG pressures mount, infrastructure design must integrate sustainability from the outset.

Smart and intelligent fields are made possible not by one single breakthrough but by the integration of many interdependent systems: sensors, networks, computing platforms, and control systems. A robust, secure, and scalable infrastructure enables oil and gas companies to move from reactive operations to proactive, predictive, and ultimately autonomous management of their assets.

3.2 DOF Asset Management

Asset management within the DOF framework represents a strategic integration of digital technologies, real-time data analytics, and intelligent systems to optimize performance and value of physical assets (Figure 3.26) [1].

It aims to extend the lifecycle of critical infrastructure while minimizing downtime, operational risk, and maintenance costs. Maintenance and integrity management are vital components in ensuring the longevity, safety, and performance of DOF technologies. As the oil and gas industry increasingly depends on advanced digital systems—such as real-time sensors, automation platforms, and analytics engines—the need to maintain and preserve the reliability of both physical and digital assets becomes critical.

In the DOF environment, asset performance and data flow are tightly linked. Failure in either hardware (e.g., sensors, control systems, communication lines) or software (e.g., data analytics, integration platforms) can disrupt the entire value chain, leading to production losses, safety risks, and increased operational costs. Therefore, maintenance and integrity strategies must evolve to include digital, predictive, and proactive approaches.

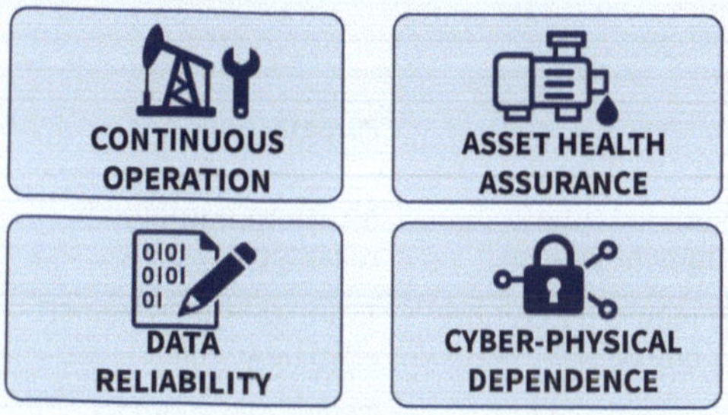

Figure 3.26 The importance of maintenance in DOF (AI generated).

There is a need to establish a maintenance contract(s) for all DOF equipment to keep them evergreen with periodic calibration and testing. Maintenance in a DOF extends beyond traditional mechanical upkeep. It encompasses a comprehensive approach to maintaining equipment functionality, system reliability, and data integrity.

3.2.1 Key Drivers for Asset Management and Maintenance in DOF

1. **Continuous Operation**: DOF systems enable real-time decision-making. Any downtime impacts field efficiency and corporate profitability.
2. **Asset Health Assurance**: Equipment such as pumps, compressors, valves, and sensors requires routine checks to prevent performance degradation and potential failure.
3. **Data Reliability**: Maintenance ensures accurate data capture and prevents misinterpretations caused by sensor drift or calibration errors.
4. **Cyber-physical Dependence**: With digital integration, maintenance must include both physical and cyber infrastructure, ensuring IT and OT systems work in harmony.

A good example of the importance of asset management can be seen in the MPFM, which needs periodic maintenance and calibration for rate measurement accuracy, without which the rate calculations that are captured, transmitted, and used by DOF workflows can lead to wrong decisions. Another example of a malfunctioning sensor or equipment that stops sending or transmitting data, resulting in several lost optimization opportunities or disasters if it is safety related. Hence, effective asset management/maintenance in DOF is therefore a strategic balance between predictive analytics, automated alerts, optimization opportunities, and human oversight, enabling DOF engineers/operators/management to intervene before issues escalate.

3.2.2 Integrity and Maintenance KPIs

To effectively monitor performance, operators may use KPIs that reflect the health, reliability, and integrity of both physical and digital assets. Tables 3.1 and 3.2

Table 3.1 Operational maintenance KPIs.

KPI	Definition	Purpose
Mean time between failures	Average operational time between equipment failures	Measures equipment reliability
Mean time to repair	Average time required to restore an asset after failure	Evaluates maintenance efficiency
Maintenance cost per barrel	Maintenance cost divided by production volume	Tracks maintenance cost-effectiveness
Planned maintenance ratio	Percentage of maintenance work that is planned versus unplanned	Assesses maintenance strategy effectiveness

Table 3.2 Integrity and reliability KPIs.

KPI	Definition	Purpose
Assets commissioning status	% of assets installed and commissioned	Monitors long-term assets' return on investment
Asset integrity index (AII DOF equipment)	Composite score reflecting the condition of critical assets	Monitors long-term asset health
Inspection compliance rate	Ratio of completed inspections to scheduled ones	Tracks adherence to integrity plans
Failure rate per asset class	Frequency of failures across asset categories	Identifies high-risk components
System availability (%)	Percentage of time the system remains operational	Reflects operational readiness
% of data transmission	Percentage of data captured, transmitted from all from sensors	Ensures reliability of data and sensors
Data accuracy rate (%)	Percentage of accurate data points from sensors	Ensures reliability of analytics

summarize the recommended operational maintenance, assess integrity, and reliability KPIs.

These KPIs are often visualized in DOF dashboards to enable continuous monitoring and strategic decision-making. Maintenance and integrity asset management form the backbone of sustainable DOF operations. Through predictive analytics, integrated KPIs, and advanced digital tools, oil and gas companies can ensure that their assets, both physical and digital, operate at optimal performance levels. A well-structured DOF asset management and maintenance strategy not only enhances asset lifespan and safety but also supports the industry's broader goals of efficiency, sustainability, and digital resilience and hence ensures the successful implementation of the distributed optical fiber technology and helps in achieving its corporate strategic objectives.

References

1 Alnuaim S. DOF and reservoir monitoring, unpublished course materials and PowerPoint presentations. KFUPM; 2020–2025.

2 Feineman DR. Assessing the Maturity of Digital Oilfield Developments. SPE Intelligent Energy International Conference and Exhibition; 2014. SPE-167832-MS.

3 International Association of Oil & Gas Producers (IOGP). Data Connectivity and Security in the Digital Oilfield; 2020.

4 Schlumberger. Digital Oilfield: Integrating Data and Control Systems; 2021.

5 International Electrotechnical Commission (IEC). Industrial Automation and Control Systems Standards. IEC 61131 and IEC 62443.
6 IEEE. Industrial Internet of Things (IIoT) and Edge Computing for Oil & Gas; 2020.
7 McKinsey & Company. Building Reliable and Secure Digital Infrastructure in Oil & Gas; 2018.
8 Oil & Gas Technology Journal. Advances in Subsea Communication Networks; 2020.
9 Deloitte. Cybersecurity Strategies for Digital Oilfields; 2021.
10 Al-Khelaiwi F, Birkeland R, Lee K. Intelligent well technology: A global review of applications. SPE Annual Technical Conference and Exhibition. Anaheim, USA; 2007. SPE-110267.
11 Yeten B, Jalali-Farahani A, Durlofsky LJ. Optimization of intelligent well completions. SPE J. 2004;9(4):458–467. SPE-84247.
12 Jalali M, Al-Hajji A. Smart well technology applications for zonal control. SPE Annual Technical Conference and Exhibition. Dallas, USA; 2000. SPE-62979.
13 Falcone G, Hewitt G, Alimonti C. Multiphase Flow Metering: Principles and Applications. Oxford: Elsevier; 2009.
14 Hartog A. An Introduction to Distributed Optical Fibre Sensors. Boca Raton: CRC Press; 2017.
15 Hoffmann R, Rønneberg H. Multiphase flow metering principles and performance. SPE Production Operations Symposium. Oklahoma, USA; 1993. SPE-25216.
16 Mohaghegh SD. Artificial Intelligence and Data-driven Reservoir Modeling. Richardson: Society of Petroleum Engineers; 2017.
17 Schlumberger. MANARA Advanced Completion System: Technical Specifications. Houston: Schlumberger; 2019.
18 Schlumberger. Optiv Fiber-optic Sensing System: DTS/DAS Overview. Houston: Schlumberger; 2021.
19 American Petroleum Institute. API 14A: Subsurface Safety Valve Systems. Washington (DC): API; 2018.
20 American Petroleum Institute. API 11S6: Intelligent Completion Equipment. Washington (DC): API; 2016.
21 International Organization for Standardization. ISO 17078: Downhole Safety Equipment. Geneva: ISO; 2020.
22 NACE International. NACE MR0175/ISO 15156: Materials for H_2S Service. Houston: NACE; 2021.
23 Al-Hashem A, Al-Azzawi M, Al-Amri K. ESP performance, failure mechanisms, and optimization. SPE Middle East Artificial Lift Conference. Muscat, Oman; 2015. SPE-177423.
24 Al-Mulhim A, Majed A. Digital ESP optimization using machine learning. SPE Artificial Lift Conference. The Woodlands, USA; 2021. SPE-204434.
25 Brown K. The Technology of Artificial Lift. Houston: Gulf Publishing; 2014.
26 American Petroleum Institute. API 21.1: Electronic Gas Metering Systems. Washington (DC): API; 2019.

27 American Petroleum Institute. API RP 86: Measurement of Multiphase Flow. Washington (DC): API; 2018.
28 Honeywell Process Solutions. SmartLine Pressure and Temperature Transmitters: Technical Guide. Phoenix (AZ): Honeywell; 2020.
29 Cameron (SLB). Adjustable and Automated Choke Valves: Engineering Data. Houston: Cameron; 2019.
30 Boyer SA. SCADA: Supervisory Control and Data Acquisition. 4th ed. Research Triangle Park (NC): ISA; 2016.
31 International Electrotechnical Commission. IEC 62443: Industrial Automation Cybersecurity. Geneva: IEC; 2018.
32 International Electrotechnical Commission. IEC 61511: Functional Safety in Process Industries. Geneva: IEC; 2016.
33 IEEE. IEEE 802.x industrial Ethernet and Wireless Communication Standards. New York: IEEE; 2020.
34 Honeywell. Experion PKS Distributed Control System Architecture. Phoenix (AZ): Honeywell; 2021.
35 Emerson. DeltaV DCS: Architecture and Control Guidelines. Austin (TX): Emerson; 2022.
36 ABB Group. ABB Ability™ System Overview for Oil & Gas Automation. Zurich: ABB; 2021.
37 Siemens. SIMATIC PCS 7 Process Control System: Technical Manual. Munich: Siemens; 2018.
38 API. API RP 1164: Pipeline SCADA Cybersecurity. Washington (DC): API; 2021.
39 Microsoft Corporation. Azure IoT Edge for Industrial Operations. Redmond (WA): Microsoft; 2022.
40 Amazon Web Services. AWS Greengrass for Edge-based Oilfield Automation. Seattle (WA): AWS; 2022.
41 McKinsey & Company. Edge Computing in Future Oilfield Operations. New York: McKinsey Energy Insights; 2021.
42 IEEE Internet of Things Journal. Edge Analytics for Hazardous Oilfield Environments. IEEE IoT J. 2017–2023.
43 Johnson A, Al-Hajri B. Real-time data conditioning and filtering in smart fields. SPE Digital Energy Conference. Houston, USA; 2018. SPE 187610.
44 OSIsoft. PI System Architecture for Oil & Gas. San Leandro (CA): OSIsoft; 2020.
45 AVEVA. Historian and Real-time Operations Platform: Technical Documentation. Cambridge (UK): AVEVA; 2021.
46 The Open Group. OSDU Data Platform: Technical Standard. San Francisco (CA): The Open Group; 2022.
47 Al-Gharbi S. Improving the Data Quality of Drilling Data, PhD Dissertation. KFUPM; 2022.
48 Mallat S. A Wavelet Tour of Signal Processing. 3rd ed. Amsterdam: Elsevier; 2009.
49 Grieves M. Digital Twin: Manufacturing Excellence through Virtual Replication. Cocoa Beach (FL): Grieves Publications; 2016.
50 IBM Corporation. Digital Twin Architectures for Heavy Industrial Assets. Armonk (NY): IBM; 2020.

51 Al-Sofi AM, Al-Hussainy S. Field-scale digital twins for production optimization. SPE Middle East Oil & Gas Show. Bahrain; 2019. SPE-196043.

52 Saudi Aramco. Journal of Technology: Sensor Systems, PDHMS, DTS, ESP, and Multiphase Metering. Dhahran: Saudi Aramco; 2005–2023.

53 Chai F, Van Den Berg F.G, Engbers P, Shell Global Solutions International B.V. Smart Fields: People, Processes and Technology; 2015. GeoExpro.com.

54 Otto G, Foreman R, Verra G. Field of the future digital infrastructure and IT architecture. SPE Intelligent Energy International Conference and Exhibition, 112149; 2008.

55 Saputelli L, Duran J, Rivas F, et al. Success cases and lessons learned after 20 years of oilfield digitalization efforts. SPE Annual Technical Conference and Exhibition; 2024.

56 API. API RP 49: H_2S Monitoring in Drilling and Production Operations. Washington (DC): API; 2019.

57 Occupational Safety and Health Administration. Hydrogen Sulfide Hazard Bulletin. Washington (DC): OSHA; 2018.

58 US Environmental Protection Agency. Methane Detection and LDAR Program Guidelines. Washington (DC): EPA; 2021.

59 Dräger. Fixed Gas Detection Systems for Oil & Gas Applications: Technical Documentation. Lübeck: Dräger; 2020.

60 Honeywell Analytics. Industrial Toxic Gas and H_2S Detection Systems. Charlotte (NC): Honeywell; 2021.

61 MSA Safety. Hazardous Gas Monitoring for Upstream Facilities. Pittsburgh (PA): MSA; 2019.

62 Temizel C, Canbaz CH, Palabiyik Y, et al. A comprehensive review of smart/intelligent oilfield technologies and applications in the oil and gas industry. SPE Middle East Oil and Gas Show and Conference; 2019. SPE-195095-MS.

63 Hassan AA, Alnuaim S. Bridging the air gap: A comprehensive workflow for integrating drone data into the digital oilfield (DOF). International Petroleum Technology Conference. Dubai; 2026. IPTC-25169-MS.

64 International Society of Automation (ISA). ISA-95: Enterprise-control System Integration. Research Triangle Park (NC): ISA; 2018.

4

DOF Financial Aspects and Business Planning

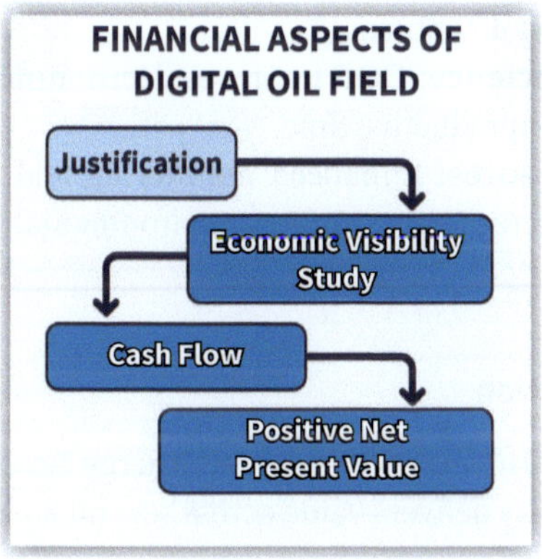

The digital oil field (DOF) is a transformative approach that leverages real-time data, automation, and analytics to optimize hydrocarbon production and asset performance. As oil and gas operators face increasing pressure to reduce operational costs and maximize recovery from aging fields, the financial justification of DOF implementation has become a central concern. A typical greenfield digitalization project cost can vary from several million dollars to several hundred million dollars based on the size, scope, and infrastructure complexity. Small fields may require \$10 M–\$100 M, while medium- to large-scale fields may require over \$100 M. Moreover, the upgrades of brown/conventional oil fields are usually more difficult and costly projects, especially in retrofitting obsolete field's scale analog systems.

This chapter explores the economic visibility, justification, and financial modeling, focusing on cash flow analysis and achieving a positive net present value (NPV) as key elements in assessing the economic visibility and viability of DOF investments.

4.1 Justification for DOF Investment

4.1.1 Industry Challenges

Traditional brown oil field operations are often fragmented, inefficient, and heavily reliant on manual, costly, and inefficient processes. DOF technology is meant to transform this inefficient oil field management environment to very efficient and optimized environment [1–6]. Key financial drivers that necessitate the adoption of DOF include the following:

- **Declining Production Rates**: Mature fields require enhanced recovery methods and real-time optimization.
- **Volatile Oil Prices**: Operators seek cost-effective technologies to maintain profitability during downturns.
- **Operational Efficiency**: DOF improves decision-making speed and reduces downtime and nonproductive time.
- **Regulatory Pressures**: Enhanced monitoring and reporting capabilities of DOF help meet increasingly stringent environmental, safety, and sustainability regulations.

4.1.2 Value Proposition

Based on the CERA 2010 study (Cambridge Energy Research Associates), a DOF implementation typically delivers value across several areas [2]:

- Improved recovery factor by 2–7%.
- Reduced capital cost by 5–30%.
- Reduce downtime by 1–2% (very conservative figure).
- Reduced operational expenditure (OPEX) cost by 5–25%.

Other recent applications of DOF showed [3, 4, 6] significant positive business impact, including:

- Increase in production.
- Decrease in maintenance costs.
- Improvement in safety, health, sustainability, and compliance.

4.1.3 Economic Visibility Study

An Economic Visibility Study (EVS) is crucial to evaluate whether DOF deployment aligns with the company's financial, production, HSE, and sustainability objectives or not. The benefits of the DOF can be easily modeled using any commercial reservoir simulator to better quantify the expected increase in ultimate recovery/production, the expected decrease in operating cost/downtime, and the expected improvement in HSE and sustainability performance. The DOF financial study incorporates the following.

4.1.3.1 Capital Expenditure

Capital expenditure (CAPEX) for a DOF varies depending on the scale, scope, and complexity of the DOF infrastructure (Chapter 3), including:

- Sensors and Internet of Things (IoT) infrastructure.
- Telemetry and communications systems.
- Supervisory control and data acquisition (SCADA) and control systems.
- IT/OT integration platforms.
- Cloud storage and cybersecurity.
- Advanced analytics and AI software.
- Relevant software, hardware, and control centers.
- DOF surface and subsurface equipment.

4.1.3.2 Operational Expenditure

Ongoing costs include:

- Software licensing.
- Data hosting and management.
- Technical support.
- Cybersecurity maintenance.
- Continuous analytics updates and staff training.

Based on experience, OPEX is often assumed to be 10–15% of the CAPEX annually.

4.1.3.3 Revenue Streams

DOF enhances revenue through:

- Increased production volume.
- Improved uptime (less deferred production) and less downtime due to equipment failure and high undesired fluids (water and/or gas), which may require rigs and expensive well treatments.
- Better recovery factor from dynamic reservoir management.
- Reduced annual OPEX over the classical project span of 20–25 years.
- Improved HSE and sustainability performance (difficult to monetize).

4.1.3.4 Cash Flow Analysis

Cash flow projections are used to assess the profitability of DOF investments over time (e.g., 20–25 years).

4.1.3.5 Assumptions for the Financial Model

Several financial assumptions can be made, depending on the operator's financial strategy and business model, which can be seen in the following example [1]:

- **Field Life**: The DOF project span could be 20–25 years. In the next example, 20 years will be used.
- Initial big investment in $ million (Year 0), example = $15 M.

- Annual production uplift revenue in $ million (due to DOF rate increase), example = $5 M.
- Annual cost savings (OPEX) in $ million (due to DOF-related operation cost savings), example = $2 M.
- Annual OPEX for DOF in $ million (due to DOF maintenance cost), example = $1.5 M.
- Discount rate: **r**, example = 10%.

4.1.3.6 Projected Cash Flows

Year	Production Uplift	Cost Savings	DOF OPEX	Net Cash Flow
0	0	0	0	−15.0 M
1–20	5.0 M/year	2.0 M/year	−1.5 M/year	+5.5 M/year

4.1.3.7 NPV Calculation

NPV helps evaluate the value created by the project in today's dollars.

$$\text{NPV} = \sum_{t=0}^{n} \frac{\text{CF}_t}{(1+r)^t} \tag{4.1}$$

where

- CF_t = cash flow in year t
- r = discount rate (10%)
- n = project duration (20 years)

Using the cash flows:

- Year 0: −15.0 M
- Years 1–20: 5.5 M annually
- Since the cash flow is the same each year, we can use the present value of an annuity formula:

$$\text{PV of Annuity} = C_t \times \left[1 - \frac{1}{(1+r)^n}\right] / r \tag{4.2}$$

$$\text{PV} = 5.5 \times \left[1 - \frac{1}{(1+0.10)^{20}}\right] / 0.10 = \$46.83 \text{ million}$$

Subtract initial investment: NPV = 46.83 − 15 = $31.83 million

4.1.3.8 Sensitivity Analysis

A sensitivity analysis helps assess the impact of key variables on NPV. Factors such as discounted rate, production uplift (could be varied either in oil price, rate, or both), and cost savings can be sensitively varied with ±% (example is ±10%). Initial investment and OPEX are usually left unchanged (more conservative approach).

4.1.3.9 Pessimistic Case: −10% Scenario and 12% Discount Rate

Year	Production Uplift	Cost Savings	DOF OPEX	Net Cash Flow
0	0	0	0	−15.0 M
1–20 years	4.5 M/year	1.8 M/year	−1.5 M/year	+4.8 M/year

NPV = $20.85. This positive NPV indicates the project is economically viable at a 12% discounted rate and −10% decrease in DOF-related production uplift and cost avoidance.

4.1.3.10 Optimistic Case: +10% Scenario and 8% Discount Rate

Year	Production Uplift	Cost Savings	DOF OPEX	Net Cash Flow
0	0	0	0	X1 = −15.0 M
1–20 years	5.5 M/year	2.2 M/year	−1.5 M/year	+6.2 M/year

NPV = $45.87. This positive NPV indicates the project is economically viable at a 8% discounted rate and +10% increase in DOF-related production uplift and cost avoidance.

Return on investment (ROI) is a financial metric used to evaluate the profitability of an investment. It measures the return relative to the investment's cost and is usually expressed as a percentage.

$$\mathrm{ROI} = \left(\frac{\text{Net Profit}}{\text{Initial Investment}}\right) \times 100 \tag{4.3}$$

where net profit = total returns or non-discounted cash flows − initial investment
Initial investment = the upfront cost of the project

- ROI (**standard case**) = (net profit/investment) × 100 = 100 × (5.5 × 20 − 15)/15 = 633%

Sometimes, special ROI is calculated based on the discounted net cash flow (NPV).

- ROI (discounted cash flow of standard case) = 100 × NPV/initial investment = 100 × (31.83/15) = 212%

The payback period is the amount of time it takes for an investment to recover its initial cost from its net cash flows. It shows how long it takes before a project starts generating profit.

$$\text{Payback Period} = \frac{\text{Initial Investment}}{\text{Annual Net Cash Flow}} \tag{4.4}$$

- Payback period (**standard case**) = initial investment/annual net cash flow = 15 M/5.5 M ≈ 2.7 years.
- Payback period (**pessimistic case**) = initial investment/annual net cash flow = 15 M/4.8 M ≈ 3.12 years.

- Payback period (**optimistic case**) = initial investment/annual net cash flow = 15 M/6.2 M ≈ 2.4 years.

4.1.3.11 Risk Considerations and Mitigation

Risks

- Technology obsolescence.
- Data security breaches.
- Resistance to change from staff.
- Vendor lock-in.
- Oil demand decreases and oil price decrease.

Mitigation Strategies

- Modular architecture for scalability.
- Robust cybersecurity policies.
- Continuous training programs.
- Open standards and interoperability.
- Operation resilience.

The NPV of the project is approximately **$31.83 million.** This positive NPV indicates the project is economically viable at a **10%** discounted rate. The project is also found to be economically viable at the **12%** pessimistic and **8%** optimistic discounted rates cases. More detailed economic feasibility studies are usually conducted based on several reservoir simulation and market conditions (supply, demand, oil price, etc.) studies.

4.2 **DOF Example** (Source: Intelligent Digital Oil and Gas Fields—Gustavo Carvajal, 2018, Permission obtained electronically from publishing company)

Field Description

- This example is found in the Ref. [1]. It describes the financial justification of a new marginal oil onshore field with 100 wells and daily production of 20 MSTBD of crude oil (30 API), 10 MSTBD of water, and 20 MMSCFD of gas, with total in situ recoverable hydrocarbon 150 MMSTB, where M = 1,000 and MM = 1000,000).
- Pressure is supported by water injections.
- Wells are lifted using ESPs.
- Oil decline rate is estimated to be 10% annually.

DOF Installation Objectives:

- Reduce the frequency of well shutdowns per month.
- Increase oil uplift and decrease oil decline.

- Delay water breakthrough and manage water production efficiently.
- Reduce development costs and operating costs.
- Increase recovery factor.
- Reduce environmental impact.

4.2.1 Cost Estimates

For an implementation plan to reach the goals stated earlier, the operator would need to equip the 100 wellheads with:

- Chokes, control panels, and actuators.
- Cables, fiber-optic cables, and routers.
- Pressure and temperature gauges.
- Sensors and other DOF equipment.
- Total cost estimate of $35 million.

Additionally, the company would need to update the data center to:

- Integrate data streams from the field to the repository center.
- Build servers, CPUs, and clusters with enough data storage for 20 years.
- Construct a collaborative working environment center.
- Generating new automated workflows and purchasing software and applications, which would cost an additional of $5 million.
- Total cost estimate of $15 million.

Grand total cost = $35 + $15 = $50 million or $500,000 per well (Figure 4.1).

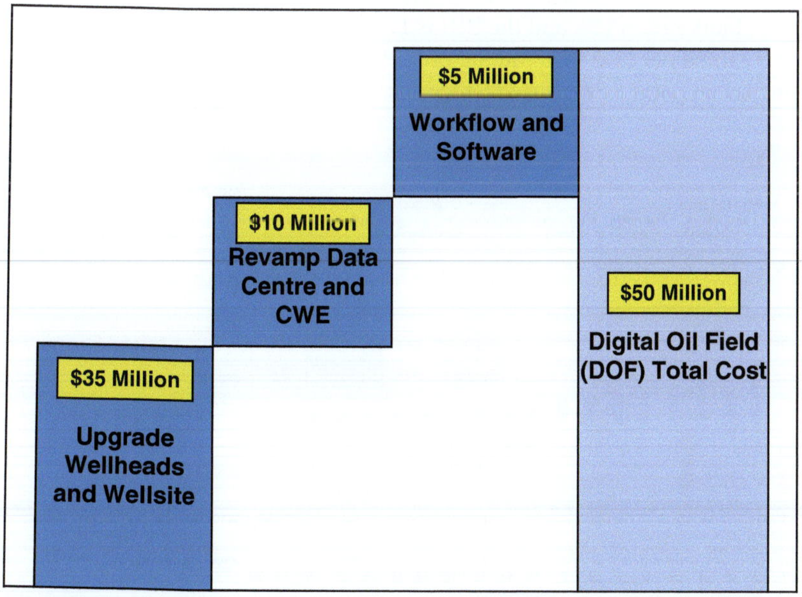

Figure 4.1 Initial capital investment for a DOF system for an onshore oil field of 100 wells (Gustavo Carvajal [1]/with permission of Elsevier).

Economic Parameters

- The economic parameters for this example include a 20-year timeframe, with an average fixed price of oil at $40 per barrel and gas at $2.75 per MSCF, a fixed interest rate of 5%, and depreciation of 10% annually (Figure 4.2).
- The operating expenses are estimated at $15 per barrel (including salary and wages).
- The revenue obtained from DOF implementation is estimated by calculating the difference between the annual oil production profiles using a DOF system compared with the production profiles without a DOF system.

Project Economic Analysis

- It is assumed that using a DOF system, ESP downtime has decreased by 70%, and also using reservoir analysis, the early water breakthrough has been delayed for more than several months by controlling the water injection. These technical assumptions of the benefits of the DOF can be easily modeled using any commercial reservoir simulator to quantify the expected increase in ultimate recovery/production and decrease in OPEX. Using the simulator and modeling the capabilities of DOF, such as ICV optimization, resulted in an increase of 12% in oil production, compared to oil production without DOF implementation.
- The initial 50-million-dollar CAPEX investment is spent in the first year. Given this analysis and an oil price of $40.0, the producer can anticipate a total benefit of $88.6 million over 20 years.
- The NPV is estimated at $64 million, and the internal rate of return is 22%, when the bank rate is 15% and the ROI is 177%.
- It is $1.7 gained per $1.0 spent on the project.
- The breakeven point is observed at 4.2 years.

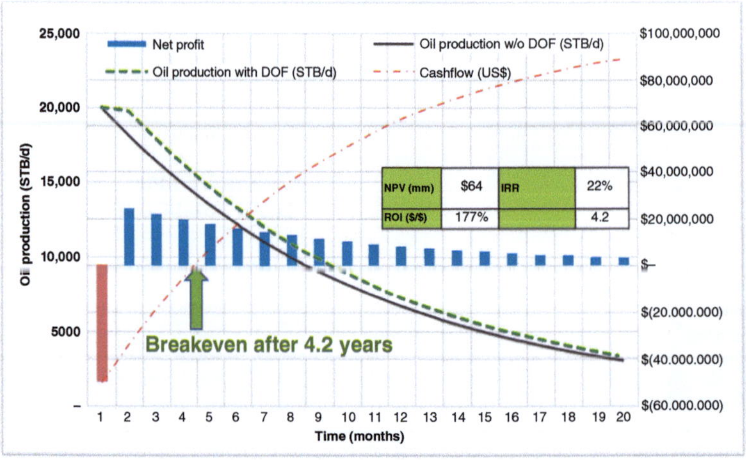

Figure 4.2 Potential ROI of a DOF implementation using a simulation study (Gustavo Carvajal [1]/with permission of Elsevier).

- The company can expect substantial economic growth in the following 16 years for a cumulative ROI of ~177%.
- DOF implementation could bring significant benefits to the company.
- Companies seek more comprehensive cost–benefit analysis considering current and future market economic and oil price forecasts.

DOF initiatives are not speculative—they are proven investments with tangible financial outcomes. International and national oil companies (IOCs & NOCs) have demonstrated positive economic visibility for most of their DOF applications, including:

- Positive NPV and ROI.
- Payback in one to three years.
- Sustainable positive gains in production, efficiency, operating cost, ultimate recovery, and HSE.

4.3 DOF Business Planning

Successful implementation of DOF demands a well-structured business plan that aligns technological capabilities with corporate strategy, economic viability, and organizational readiness [3–6].

The first step in planning a DOF initiative is defining its strategic intent within the company's broader energy and technology roadmap. Common objectives include:

- **Production Optimization**: Maximizing output through predictive analytics and automated control.
- **Operational Efficiency**: Reducing downtime, maintenance costs, and human error.
- **Safety and Compliance**: Leveraging real-time monitoring to improve worker and environmental safety.
- **Sustainability**: Reducing carbon footprint via energy-efficient operations.
- **Data-driven Decision-making**: Enabling management to respond rapidly to field dynamics.

The vision should clearly connect digital transformation to corporate business outcomes, not just technology adoption. A robust financial plan ensures sustainable implementation and scalability.

Investment Components

- **CAPEX**: Hardware (sensors, PLCs, SCADA), infrastructure upgrades, and communication networks. CAPEX often accounts for 40–60% of the total initial investment, which includes:
  - **Hardware**: Industrial sensors, PLCs, RTUs, drones, communication towers, edge devices.

  - **Control Systems**: SCADA, distributed control systems (DCSs), and automation servers.
  - **Network Infrastructure**: Fiber-optic networks, satellite links, private 5G, or secure IoT gateways.
  - **Data Centers and Cloud Migration**: Computing and storage capacity for high-volume real-time data.
  - **Integration Projects**: Engineering, installation, and commissioning of the digital stack.
- **Operational Expenditure (OPEX)**: Software licenses, data analytics subscriptions, maintenance, and workforce training. Recurring costs that maintain and operate the digital ecosystem, including:
  - **Software Licensing and Subscriptions**: Cloud services, analytics dashboards, cybersecurity platforms.
  - **Maintenance and Support**: Hardware calibration, software updates, and vendor service contracts.
  - **Training and Workforce Development**: Upskilling technicians, data analysts, and control engineers.
  - **Energy and Connectivity Costs**: Power supply for remote equipment and data transfer.
  - **Data Governance and Security Monitoring**: Ongoing cyber-resilience measures.
- Hidden and indirect digital transformations also incur non-technical expenditures that must be forecasted:
  - Change management programs and organizational restructuring.
  - Business process redesign to align workflows with digital systems.
  - Consultancy and compliance costs (e.g., ISO 27001 certification, data protection audits).
- **Cost Savings Projections**: Estimates from reduced downtime, predictive maintenance, and optimized drilling operations.
- **ROI and Payback Period**: Typically, three to five years, depending on field maturity and scope.

Failing to account for these often leads to underestimated budgets and delayed ROI.

Funding models may include direct investment, joint ventures, or technology partnerships with service providers.

To ensure continuous improvement, the DOF business plan must define key performance indicators (KPIs) such as:

- Production efficiency (barrels per day per well).
- Downtime reduction (%).
- Maintenance cost savings (%).
- Data accuracy and latency.
- Environmental impact metrics.

Analytics dashboards can visualize KPI trends and facilitate real-time decision-making.

Key risk categories to be included in the DOF business plan include:

- **Technical Risks**: System integration issues or sensor failures.
- **Cybersecurity Risks**: Threats to connected infrastructure.
- **Financial Risks**: Cost overruns or delayed ROI.
- **Oil Price**: Fluctuation of the oil price during the project life span
- **Human Factors**: Resistance to technological change or lack of skills.

Aligning CAPEX and OPEX structures with a robust risk mitigation framework and contingency funding transforms a DOF project from a technological experiment into a strategic investment portfolio.

Finally, while DOF projects require upfront capital and strategic alignment, the financial rewards make the DOF a financially compelling investment for upstream operators. With a solid business case, positive NPV, rapid short payback period, and significant ROI, the economic rationale for digital transformation is strong. When implemented strategically and supported by robust risk mitigation, skilled staff, and change management, DOF initiatives can serve as a sustainable catalyst for value creation in a volatile energy landscape. In the past, the oil and gas industry had to justify any DOF investment with extensive economic visibility and risk studies. Today, any major upstream new development projects undertaken by major NOCs and/or IOCs have a DOF arm by default. Hence, it is becoming part of any upstream business and operating plans.

References

1 Carvajal G. Intelligent Digital Oil and Gas Fields. Houston: Oilfield Books, Elsevier; 2018.

2 Cambridge Energy Research Associates (CERA). The Digital Oilfield of the Future: Enabling Next-generation Reservoir Performance. Cambridge (MA): CERA; 2010.

3 Deloitte. Digital Oilfield: The Upstream Industry's New Horizon. London: Deloitte Insights; 2018.

4 Accenture. Realizing the Value from Digital Oilfield Transformations. Dublin: Accenture Energy; 2019.

5 Alnuaim S. DOF and reservoir monitoring, unpublished course materials and PowerPoint presentations. KFUPM; 2020–2025.

6 Schlumberger. Digital Oilfield Economics: Real-time Operations and Financial Impact. Houston: Schlumberger; 2020.

5

DOF Organizational Change and Workforce Transformation

Empowering People for the Intelligent Energy Era

Technology may power smart and intelligent fields, but skilled and talented people make them possible. While the benefits of digital transformation in oil and gas fields are well-documented, including higher productivity, safer operations, and lower environmental impact, these gains cannot be fully realized without parallel shifts in organizational structure, culture, skills, and leadership. Smart field implementation is not merely an IT or engineering project; it is a business wide transformation that touches every layer of the organization. It requires new ways of working, fresh mindsets, and the ability to learn, adapt, innovate, and collaborate across traditional boundaries using untraditional methods and very advanced technologies.

In this chapter, we will explore how oil and gas companies are rethinking their workforce strategies and organizational models to thrive in the digital era. From talent development and innovation to change management, this is where the human side of the intelligent oil field comes into focus. The structure of the chapter will be based on my understanding of the effective digital organizational/people transformation journey that took place in the last two decades, summarizing a rich literature, publicly published documents, and case studies [1–33].

5.1 Importance of Organizational Transformation

Deploying digital infrastructure and analytics platforms without organizational change leads to:

- Underutilized technologies due to a lack of adoption and DOF advantages.
- Resistance from field teams rooted in the fear of job loss or lack of trust.
- Data silos and process fragmentation across departments. Conventional data access policies hinder the harvest of DOF objectives; hence, an agile data access strategy has to be developed and implemented with best-in-class data management, data access, and data security practices, possibly using AI and ML advanced and smart algorithms.
- Misalignment between technology capabilities and business goals.

To avoid these pitfalls, companies must align their digital transformation strategy with a people-first strategy that empowers and continuously trains employees (engineers, scientists, operators, team leaders, management, and executive management) to adopt and sustainably shape smart field innovations.

5.2 Cultural Shifts in Smart Field Organizations

The digital oil field demands a culture of agility, experimentation, and continuous learning. Key cultural attributes include:

- **Data-driven Decision-making**: Replacing intuition-based choices with evidence, advanced analytics, machine learning (ML), and artificial intelligence (AI).
- **Cross-functional Collaboration**: Breaking down silos between subsurface, operations, IT, and business teams. The concept of digital assets' teams is an effective organizational business model, where the DOF asset is managed by highly skilled and digitally enabled multidisciplinary teams of engineers, geoscientists, data analysts, and empowered management.
- **Innovation Mindset**: Encouraging pilots, failures, and iteration as part of the learning process. Establishing innovation centers equipped with DOF simulation capabilities for training, trial/errors, and pilots' studies will positively enrich the organization's digital culture shift and maturity.
- **Transparency and Trust**: Ensuring all levels of staff understand the purpose and impact of smart field technologies through full alignment of the DOF, organization, and corporate strategies.
- **Sustainability Awareness**: Embedding ESG considerations into day-to-day decision-making through smart and intelligent digital solutions.
- Strong data management and data access strategy to avoid data silos and maximize the utilization of the DOF data streams.
- Strong IT infrastructure and digitization maturity, focusing more on the human factor, organizational structure, and change management practices.

Leaders from the top of the organization pyramid to the first-line management to the lower-field engineers/operators must serve as champions of these shifts, modeling the behavior they wish to see across the organization [1–33].

5.3 Evolving Roles in the Digital Oil Field

As automation, AI, and digital tools become more dominant, workforce roles of traditional oil and gas companies are being redefined [1–33]. Key workforce shifts include the following (Table 5.1).

In many cases, these roles require hybrid skill sets of blending engineering, data science, analytics, smart software, AI/ML, and business acumen, without which the DOF benefits may not be effectively achieved. The success of DOF implementation will be directly proportional to the maturity of this factor in each corporation.

5.4 Workforce Development and Digital Upskilling

To equip personnel for smart field operations, companies are investing in comprehensive upskilling and reskilling programs for employees, including:

- Digital academies offering courses in data science, AI, ML, cloud computing, and automation. Official accreditations in these sub-DOF domains can be achieved through structured programs offered by several professional, educational, and vocational training organizations.
- Simulation-based training for remote and AI-enhanced decision-making.
- Cross-disciplinary learning between engineers and IT professionals.
- On-the-job digital mentoring programs for real-world experience.
- Macro (data analytics) and micro (AI) accreditation and certification strategy/program to track progress and competencies of all levels of DOF technical and management staff. This also includes cybersecurity accreditation programs.

Table 5.1 A typical DOF workforce role comparison (traditional and emerging).

Traditional Role	Emerging Role
Classical drilling, production, facility, operation and reservoir, engineers	**1.** Digital skilled drilling, production, facility, operation and reservoir engineer with modeling and data analysis **2.** Upstream data scientist: upstream data modeler
Control room operator	**1.** Remote operations analyst **2.** DOF engineer
Maintenance technician	Predictive maintenance specialist
Geoscientist	AI-augmented subsurface analyst
IT specialist	OT/IT/IoT/AI/ML convergence trained consultant
Field supervisor	Intelligent mobile-enabled field coordinator

- Collaboration business models among IOCs, NOCs, and major service providers of AI/ML/IoT/data analytics/cloud computing, such as Microsoft, Intel, Oracle, and Google.

Digital technology is an evolving technology with a significantly fast pace of technological uplift. Hence, DOF upskilling is not a one-time event but a continuous process aligned with evolving technologies, field realities, and operational challenges [1–33].

5.5 Change Management Strategies

Digital transformation success depends heavily on effective change management. Key components include:

- Clear communication of vision and benefits tailored to different employee levels (linked to corporate vision, strategic goals, and incentive and performance programs for employee/management).
- Engagement of frontline workers in pilot programs and co-design of digital tools.
- Quick wins and success stories to build momentum and internal support.
- Change agent or "digital champion" embedded in each operational team.
- Feedback loops to iteratively improve digital adoption and usability.
- Clear documented/communicated roles and responsibilities as well as accountability at all organization levels.

Resistance to change is a natural, but it can be reduced through involvement, education, incentive programs, and more focus on solving real problems, especially quick wins [1–33].

5.6 Leadership in the Age of Intelligent Fields

Digital transformation requires a new kind of leadership, one that is technically informed, adaptive, digital-people-centric, and emotionally intelligent. Attributes of effective leaders in smart field organizations include:

- **Visionary Thinking**: Aligning smart field investments with long-term organization/corporate business and sustainability goals.
- **Digital Fluency**: Understanding the basics of AI, data, and digital platforms.
- **Empowerment over Control**: Trusting teams and providing them the tools to innovate, pilot, trial/error, and learn from failures.
- **Cross-functional Leadership**: Leading across departments, functions, and geographies, especially in IOCs, because they operate in wide geographical areas where emotional intelligence is essential.
- **Agility and Resilience**: Navigating change and uncertainty with composure.

Corporate leadership development programs must now include several modules focusing on management/leadership digital literacy and transformational leadership [1–33].

5.7 Diversity, Inclusion, and the Digital Workforce

As the industry evolves, there is an opportunity to build a more inclusive and diverse workforce, especially in digital roles:

- **Broader Talent Pipelines**: Attracting candidates from data science, computer science, and sustainability backgrounds.
- **Gender and Cultural Diversity**: Improving representation in technical and leadership roles.
- **Global Collaboration**: Leveraging talent across borders through digital tools, strong connectivity, global public/private clouds, and geographic talent resources (e.g., India and Silicon Valley).
- **Remote Work Policies**: Enabling broader participation in field operations without geographic constraints through virtual and overseas offices and out-of-the-box business models.
- **Part-time Work Policies**: Enabling the industry for cheap and fast quick wins using freelancers and college students' relatively cheap resources.

Inclusion is not only ethical—it improves innovation and decision-making in a very complex energy, data-driven environment [1–33].

5.8 Organizational Structures for Smart Field Integration

To support smart operations, companies must restructure their organizational models, including:

- Digital centers of excellence (DCoEs) to centralize innovation, governance, and best practices.
- Integrated operations centers (IOCs) combining real-time data, remote operations, and decision-making under one roof.
- Cross-functional agile teams focused on specific use cases or asset clusters.
- Dual-career tracks for engineers, geoscientists, business, and digital specialists to grow without leaving their domain.
- Hybrid work environments blending field presence with remote digital collaboration and simulation environment.

Organizational agility and flexibility are essential to respond quickly to technological advances and market dynamics [1–33].

5.9 Multidisciplinary Asset Team/Management in the DOF

In a modern DOF environment, multidisciplinary asset teams work together within a centralized control room equipped with real-time data visualization and digital twin technologies and complemented with the best-in-class field, reservoir, HSE, and sustainability workflows (Figure 5.1) [1–33].

These integrated teams drive fast, evidence-based decisions, reduce operational silos, and enhance overall field performance. A typical DOF asset management structure comprises experts from all upstream disciplines working in harmony to manage the asset's life cycle with the following DOF three main objectives: increasing production/recovery, reducing development cost ($/barrel), and significantly improving HSE/sustainability performances.

5.9.1 Core Team Composition

Table 5.2 shows a typical DOF core team composition, functions, and key responsibilities (it may vary from one company to another based on the specific company organizational structure and business model).

The central control room (CCR) serves as the nerve center of the DOF core team.

5.9.2 CCR Key Features

- **Unified Visualization Dashboards**: Consolidate well, reservoir, and surface data in real time.

Figure 5.1 A typical DOF multidisciplinary asset team (AI generated).

Table 5.2 A typical DOF core team composition.

Function	Key Responsibilities in DOF Operations
Reservoir engineer	Analyze subsurface models, forecast production, and optimize reservoir drainage using real-time well test, real-time well performance, digital twins, and reservoir simulation studies
Production engineer	Monitor well and surface network integrity and performance, manage choke optimization, and analyze lift efficiency and performance (e.g., Gas Lift and ESP)
Drilling engineer	Manage real-time drilling data, trajectory optimization, geo-steering, and rig performance monitoring
Geoscientist	Update geological models and integrate seismic data with production and drilling outcomes. Participate in the geo-steering function with real-time updates of the geological model based on real-time data
Facilities engineer	Oversee real-time surface equipment performance, flow assurance, and separation/compression systems, and conduct real-time network debottlenecking studies of surface/subsurface facilities
Data scientist and automation specialist	Develop data analytics models, predictive algorithms, and automated intelligent workflows (e.g., ESP efficiency and failure prediction and smart well integrity forecast)
IT, AI, ML expert	Develop multitechnical domain AL and ML advanced algorithms with prediction capability
Operation supervisor	Coordinate real-time field execution and provide feedback to overall field digital models for continuous improvement

- **Digital Twin Integration**: Dynamic surface/subsurface 3D models simulate field conditions to assess the impact of operational changes before implementation and for optimization/debottlenecking after implementation.
- **Collaborative Workflows**: Engineers and analysts review system alerts, run simulations, and execute optimized operational actions jointly.
- **Decision Support Systems**: AI-driven analytics provide automated insights, suggesting best operational actions based on multiple scenarios.
- Equipped with real-time smart field, reservoir, HSE, and sustainability workflows.

5.9.3 Advantages of Integrated Decision-Making:

1. **Speed**: Decisions that once took days can now be made in real time.
2. **Accuracy**: Cross-disciplinary validation reduces uncertainty in interpretation.
3. **Optimization**: Combined use of surface and subsurface data allows full-system optimization, from the reservoir to the export pipeline, maximizing optimization and efficiency.
4. **Risk Reduction**: Predictive modeling and simulation minimize trial-and-error interventions.
5. **Cost Efficiency**: Coordinated actions reduce redundant efforts and optimize maintenance schedules.

For instance, a reservoir engineer analyzing pressure data can instantly collaborate with a production engineer to adjust choke settings, using a digital twin simulation to test the change virtually before applying it in the field. Strategic benefits of multidisciplinary DOF teams/management include: enhanced field recovery, reduced production downtime, improved safety, lowered OPED, and increased reporting transparency. The multidisciplinary asset team model is the foundation of successful digital oil field operations. By fostering collaboration across engineering disciplines and leveraging advanced tools like digital twins and real-time analytics, organizations achieve faster, more informed decisions that enhance production, reduce costs, and extend asset life.

5.10 Future Outlook: The Intelligent Workforce

As AI and autonomous systems continue to evolve, the future oil and gas workforce will:

- Focus on oversight, strategy, and exception management rather than manual tasks.
- Collaborate with AI systems to improve outcomes in drilling, reservoir management, and logistics.
- Continuously learn and adapt, supported by real-time feedback and training platforms.
- Work more safely, with robots and drones handling hazardous tasks.
- Play a central role in sustainability, using digital tools to reduce environmental impacts.
- Timely and skillfully deal with real-time operational issues, including their integration with drones and robots, especially as the industry moves forward toward fully automated unmanned intelligent operations.

It is very important to know that the smart digital oil field is not about replacing people—it's about augmenting their capabilities and elevating their skills/roles. Digital transformation in oil and gas is not only a technological evolution but a human revolution. For smart and intelligent fields to succeed, companies must invest equally in their people and culture as they do in platforms and systems. The next generation of energy engineers/professionals will not just operate wells and facilities. They will manage data, algorithms, and AI-powered ecosystems. Preparing them for this future is not optional; it is essential for long-term success and resilience [1–33].

In the next chapter, we turn our attention to other critical elements of the digital transformation journey: cybersecurity, data governance, business continuity, and disaster recovery.

References

1 Carvajal G. Intelligent Digital Oil and Gas Fields. Houston: Oilfield Books, Elsevier; 2018.
2 Deloitte. Digital Transformation in Oil & Gas: Impact on Workforce and Organizational Agility. London: Deloitte Insights; 2020.
3 McKinsey & Company. The Human Factor in Digital Transformation: Skills, Culture, and Change Leadership. New York: McKinsey Global Institute; 2019.
4 Accenture. Workforce Transformation in the Era of Intelligent Operations. Dublin: Accenture Energy; 2021.
5 IBM Institute for Business Value. AI and the Future of Work in Industrial Sectors. Armonk (NY): IBM; 2020.
6 PwC. Workforce of the Future: The Competing Forces Shaping 2030. London: PricewaterhouseCoopers; 2018.
7 Korn Ferry. Digital Leadership Competencies for Future-ready Organizations. Los Angeles (CA): Korn Ferry Institute; 2019.
8 Schein EH. Organizational Culture and Leadership. 5th ed. Hoboken (NJ): Wiley; 2017.
9 Kotter JP. Leading Change. Boston (MA): Harvard Business School Press; 2012.
10 Harvard Business Review. The Culture Factor in Digital Transformation. Boston (MA): HBR Press; 2018.
11 Society of Petroleum Engineers (SPE). Building digital competencies in upstream engineering teams. SPE Intelligent Energy Conference. Aberdeen, UK; 2019. SPE-194567.
12 Al-Dhubaib T. Digital competency development for smart field workforce. SPE Middle East Oil & Gas Show. Bahrain; 2021. SPE-205492.
13 Al-Sofi AM, Al-Hussainy S. Integrated operations centers and multidisciplinary workflow transformation. SPE Annual Technical Conference. Dallas, USA; 2018. SPE-191519.
14 Schlumberger. Digital Competency and Remote Operations Training Framework. Houston: Schlumberger; 2020.
15 Microsoft. Digital Skilling for Energy Professionals: AI, Cloud, Data and Cybersecurity Pathways. Redmond (WA): Microsoft; 2022.
16 Google Cloud. AI and ML Learning Pathways for Industrial Workforce Transformation. Mountain View (CA): Google; 2021.
17 AWS. Industrial Workforce Digital Upskilling using Cloud-native Tools. Seattle (WA): Amazon Web Services; 2022.
18 Society of Human Resource Management (SHRM). Managing Organizational Change in Technology-driven Industries. Alexandria (VA): SHRM; 2019.
19 International Labour Organization (ILO). The Future of Work in Energy: Digitalization, Automation, and Workforce Resilience. Geneva: ILO; 2022.
20 MIT Sloan Management Review. Leading Digital: Turning Technology into Transformation. Cambridge (MA): MIT Press; 2019.

21 Gartner. Digital Dexterity and Future Workforce Readiness. Stamford (CT): Gartner Research; 2021.

22 Harvard Kennedy School. Leadership for a Digital World: Skills and Behaviors for Modern Executives. Cambridge (MA): Harvard University; 2020.

23 EY Global Oil & Gas. Operating Models and Organizational Design for Digital Upstream Companies. London: Ernst & Young; 2021.

24 Boston Dynamics. Robotics Integration and Workforce Safety in Industrial Operations. Waltham (MA): Boston Dynamics; 2022.

25 International Energy Agency (IEA). Digitalization & the Future Energy Workforce. Paris: IEA; 2021.

26 Bain & Company. Organizational Agility and Human Capital in Energy Digital Transformation. Boston (MA): Bain Insights; 2020.

27 World Economic Forum. The Future of Jobs: Skills Evolution, Automation Trends, and Digital Roles. Geneva: WEF; 2023.

28 National Academies of Sciences. AI, Workforce Development, and Industrial Transformation. Washington (DC): NAS Press; 2021.

29 O'Reilly T. Empowered: Ordinary People, Extraordinary Digital Transformation. San Francisco (CA): O'Reilly Media; 2020.

30 IEEE Internet of Things Journal. Human–machine Collaboration in Digital Industrial Operations. New York: IEEE; 2019.

31 Chai CF, Van Den Berg FG, Engbers P. Shell Global Solutions, Smart Fields: People, Processes and Technology. December 8, 2015, GeoExPro.com.

32 Temizel C, Canbaz CH, Palabiyik Y, et al. A comprehensive review of smart/intelligent oilfield technologies and applications in the oil and gas industry. SPE Middle East Oil and Gas Show and Conference; 2019. SPE-195095-MS.

33 Saputelli L, Duran J, Rivas F, et al. Success cases and lessons learned after 20 years of oilfield digitalization efforts. SPE Annual Technical Conference and Exhibition; 2024.

6

DOF Cybersecurity, Business Continuity, and Disaster Recovery

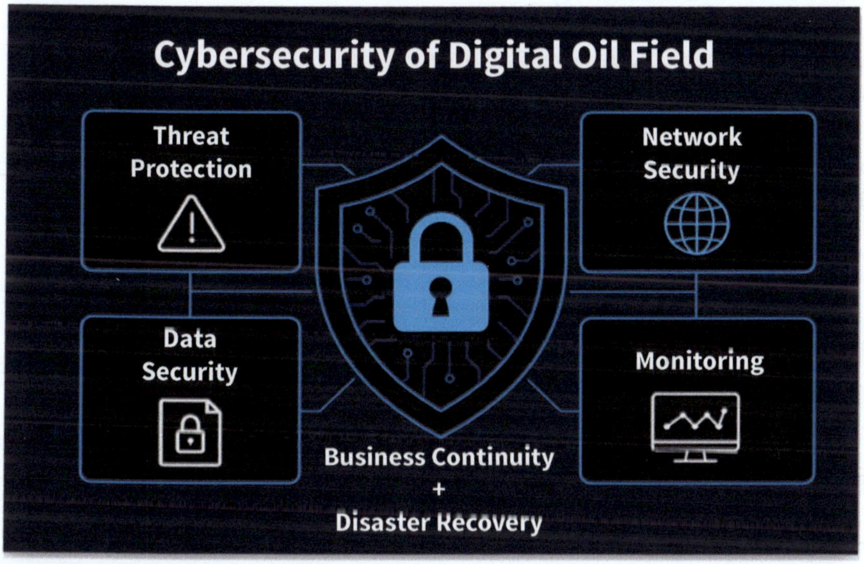

Availing and Safeguarding the Digital Oil Field

As oil and gas fields become increasingly digitized, connected, and autonomous, cybersecurity, business continuity, and disaster recovery (DR) have emerged as core pillars of smart and intelligent field success. The same technologies that enabled real-time optimization—Internet of Things (IoT) devices, cloud platforms, AI algorithms, and remote operations—also introduced new vulnerabilities that can compromise safety, disrupt operations, and damage reputations. Cyber threats are no longer theoretical. From ransomware attacks targeting refineries to malware infiltrating pipeline control systems, the energy sector has become a prime target for cybercriminals and industrial espionage. At the same time, as digital oil field (DOF) systems become increasingly integrated and data-intensive, maintaining high system availability and ensuring robust DR capabilities are becoming mission-critical.

This chapter explores the need and the how for oil and gas companies to secure their digital infrastructure and build a strong foundation for data trust—ensuring that the smart field is not only a good investment but also resilient, ethical, and smartly secured using state-of-the-art international cybersecurity standards. Furthermore, this chapter explores the key principles, challenges, strategies, and best practices for achieving reliable system availability and DR [1–18].

6.1 DOF Cybersecurity

No effective and empowered oversight of cybersecurity organization presence could hinder the whole DOF technology and subject corporations to frequent attacks and system/production interruptions. In addition, securing DOF environments requires specialized tools such as industrial firewalls, deep packet inspection, and passive network monitoring tailored to supervisory control and data acquisition (SCADA), distributed control systems, and programmable logic controller systems [1, 5, 6, 8, 17–20].

6.1.1 The Rising Threat Landscape in Oil and Gas

The convergence of information technology (IT), operational technology (OT), and IoT-connected/integrated systems has introduced new threats. Key cybersecurity risks include:

- Malware and ransomware disabling control systems or encrypting operational data.
- Unauthorized access to SCADA systems or field devices.
- Data interception through unsecured communication channels.
- Insider threats from employees or contractors with access to critical systems.
- Supply chain attacks through third-party software or hardware components.
- Disruption of remote operations due to compromised VPNs or cloud services.
- Data leaks of supercritical and confidential information.

With the growth of remote monitoring, cloud integration, and edge computing, the smart field is more exposed than ever.

6.1.2 Key Principles of Industrial Cybersecurity

To defend against these threats, organizations must implement a defense-in-depth strategy. Cybersecurity defense strategy core principles include:

- **Zero Trust Architecture**: Assume no device or user is trusted by default; verify everything.

- **Network Segmentation**: Isolate IT and OT networks to limit attack spread.
- **Least Privilege Access**: Ensure users only have access to the data and systems required for their roles. This must be kept evergreen since people move across organizations.
- **Data Encryption**: Protect data in transit and at rest to prevent unauthorized interception.
- **Intelligent Monitoring**: Continuous monitoring and anomaly detection to identify threats in real time. Use smart AI, ML, and advanced analytics.
- **Incident Response Planning**: To effectively respond to contain, mitigate, and recover from any cyber events with no core business interruption.
- **Authentication**: The use of multi-authentication/identification systems and complex password strategies. Awareness and drill campaigns are essential.
- **Drills**: Continuous cybersecurity drills with organization/individual performance monitoring, which must be linked to organizational performance and employees' incentive programs.
- **Cybersecurity Culture Measures**: Focus on organizational cybersecurity culture with KPIs and measurement systems.
- **Change Management**: Continuous cybersecurity change management practices using the best-in-class tools, processes, and technologies.
- **Oversight Organization**: Establishing a separate and empowered cybersecurity organization to have oversight and a governing role over the entire corporation.
- **Third-Party Certification**: All corporate cybersecurity functions must be certified by a reliable cybersecurity third-party organization (with no conflict-of-interest relationship with the corporation).

These practices must be embedded within each corporation with drill-down into every layer of smart field infrastructure—from sensors to cloud platforms.

6.1.3 Cybersecurity Standards and Frameworks

Leading companies adopt internationally recognized frameworks to guide their cybersecurity strategies, including:

- **NIST Cybersecurity Framework (CSF)**: Comprehensive guidelines for identifying, protecting, detecting, responding to, and recovering from cyber threats.
- **ISA/IEC 62443**: Security for industrial automation and control systems.
- **ISO/IEC 27001**: Information security management systems.
- **OG86 and API 1164**: Energy industry-specific guidelines for pipeline and control system cybersecurity.

Details of these standards can be found on these organizations' websites [1–5]. Compliance with these standards is critical not only for internal resilience but also for regulatory audits and stakeholder assurance. This is in addition to the effective cybersecurity change management with frequent systems updates and announced/unannounced drills [1–5, 19, 20].

6.1.4 Data Access in Smart Fields

Beyond protection, companies must manage operational data as a strategic asset. It is very important to create an organizational culture that considers upstream data as valuable as oil [10, 12, 14]. Effective data governance includes:

- **Data Ownership and Stewardship**: Defining who is responsible for data quality, access, and usage.
- **Data Lineage**: Tracking where data comes from, how it changes, and where it's used.
- **Metadata Management**: Cataloging datasets to ensure clarity and traceability.
- **Master Data Management**: Standardizing core data entities such as wells, assets, and equipment.
- **Data Privacy and Compliance**: Adhering to data regulations or local data sovereignty laws (e.g., Ministry of Energy in case of NOCs).
- **Data Access Policy**: Each company has its own data access policy, which must be flexible and secure to harvest the fruits of the DOF benefits.
- **Data Life Cycle Policy**: Each data must be governed and classified with a specific life cycle, from generation all the way to final permanent storage or deletion.
- **Data Leak Policy**: A data leak policy for a DOF application is a formal set of security rules, procedures, and technologies designed to prevent the unauthorized exposure, transmission, or access of sensitive data from within the DOF environment to external or unauthorized destinations.
- Effective data governance ensures accuracy, consistency, and reliability of information across the smart field and maximizes the benefits of DOF.

6.1.5 Cloud and Edge Security

Smart fields increasingly rely on cloud computing and edge devices to process and store data. Both require robust security measures such as:

- Role-based access controls and multifactor authentication.
- Secure APIs and encrypted data transfers.
- Data segregation for multitenant environments.
- Continuous vulnerability assessments and patching.
- Backup and DR protocols.
- Physical access controls,
- Local firewalls and intrusion detection.
- Remote management with secure authentication.

These safeguards ensure data integrity and system availability even in distributed environments. More details can be found in relevant references [9, 10, 17].

6.1.6 Third-party and Supply Chain Risk

Smart field ecosystems often involve numerous vendors for sensors, software, networking, and cloud services. This creates a shared security responsibility:

- Vendor risk assessments should be conducted before onboarding any supplier.
- Security service level agreement (SLA)and contract clauses must specify responsibilities, breach reporting, and compliance obligations.
- Software contracts must contain clauses for tracking and monitoring vulnerabilities, including in third-party code.
- Ongoing audits and penetration testing to verify security claims. Third-party certification may also be necessary.

Supply chain resilience is now a key component of the DOF cybersecurity strategy.

6.1.7 Building a Cyber-resilient Culture

Technology alone cannot ensure security. Organizations must cultivate a cyber-aware culture [1–18]. The following are good examples:

- Employees' training and phishing simulations. Employee's incentive and annual performance programs must be linked to his/her annual cybersecurity performance.
- Incident drills and exercises to test response capabilities.
- Executive engagement to prioritize cybersecurity at the board level.
- Clear policies and consequences for misuse or negligence.
- Collaboration with industry peers and government to share intelligence and best practices.

6.2 DOF Business Continuity and DR

The DOF represents the convergence of OT and IT to optimize the exploration, production, and management of hydrocarbon assets. As DOF systems become increasingly integrated and data-intensive, maintaining high (~100%) system availability and ensuring robust DR capabilities are becoming corporate mission-super-critical. This section explores the key principles, challenges, strategies, and best practices for achieving reliable system availability and DR in a typical generic DOF environment [1–18].

6.2.1 Importance of Availability in DOF Systems

- **Business Continuity**: Downtime in DOF systems can halt real-time monitoring, predictive analytics, and remote operations, leading to operational delays, production losses, safety risks, and financial losses.

- **Safety and Compliance**: High availability supports safety-critical systems such as well control, pressure monitoring, environmental protection, and sustainability performance, ensuring compliance with industry regulations (e.g., API, ISO, IEC, SPE).
- **Real-time Decision Support**: Live data feeds from sensors, SCADA, historians, and analytics tools enable timely decisions. Availability ensures uninterrupted data flow and situational awareness.

6.2.2 Threats to System Availability

- **Hardware Failures**: Servers, sensors, network devices, satellites, third-party IT services, and storage units can fail, affecting field data acquisition and processing.
- **Software Failures**: Bugs, misconfigurations, or integration issues can disrupt DOF platforms, including SCADA, E&P software, and databases.
- **Cybersecurity Attacks**: Malware, ransomware, or distributed denial-of-service attacks targeting critical infrastructure can bring down entire systems.
- **Environmental Disasters**: Earthquakes, floods, hurricanes, fires, or storms may damage data centers, control rooms, and field infrastructure.
- **Physical Security**: Physical security is also super critical, especially in tension and war geographical areas.

6.2.3 High Availability Strategies

- **Redundancy**: Hardware Redundancy: Dual power supplies, RAID storage, redundant network paths. System Redundancy: Active-active and active-passive clusters, especially for critical components like SCADA servers or OPC interfaces [1–18].
- **Load Balancing**: Distributes workloads across multiple systems to avoid overloading and single points of failure.
- **Fault Tolerant Architectures**: Implementing designs that allow systems to continue operation despite component failures, such as real-time failover.

6.2.4 Monitoring and Alerting

24/7 health checks, alerts, and predictive maintenance using reliable vendor tools or other customized proprietary oil field solutions with AI, ML, and advanced algorithms.

6.2.5 DR Planning

Recovery Objectives

- Maximum acceptable downtime.
- Maximum acceptable data loss window.

DR Sites and Replication

- **Cold Site**: Backup infrastructure that needs manual activation.
- **Warm Site**: Partially active, faster recovery than cold sites.
- **Hot Site**: Fully operational standby systems with real-time replication, especially for supercritical operations.

Backup Strategies

- Full, incremental, and differential backups.
- On-site and off-site storage, including cloud-based DR-as-a-service.

6.2.5.1 DR Drills and Testing

Regular simulation of DR scenarios to validate response plans, staff readiness, and infrastructure resilience.

6.2.6 Best Practices for DOF Availability and DR

- Conduct risk assessments to identify critical assets and vulnerabilities.
- Implement tiered recovery plans based on business impact.
- Establish SLAs with vendors for support and uptime.
- Regularly test DR plans to ensure effectiveness.
- Document procedures for failover, backup, and restoration.
- Train personnel on emergency protocols and system recovery steps.
- Adopt cybersecurity best practices (e.g., firewalls, patching, and access control).

6.2.7 DOF System Recovery After a Cyber Incident

Operators who experience a ransomware attack that disables their strategic real-time operations (e.g., drilling) should have the capabilities to restore full operational capabilities within a few hours if they have a strong and effective business continuity and DR strategy, including:

- Regular continuously update off-site backups.
- A hot standby environment in a remote data center recommended for all operators' strategic business processes.
- Predefined DR procedures and response plans.

6.2.8 Future DR Trends

- **AI-driven Resilience**: Using machine learning for predictive DR and self-healing systems.
- **Blockchain for Audit Trails**: Immutable logs for compliance and disaster investigations.

- **Autonomous DR**: Cloud-native solutions that detect and recover from failures without human intervention.
- **Zero Trust Security Models**: Minimizing breach impact in hybrid DOF environments.

6.3 Cybersecurity in Action: Examples of Energy Real-world Incidents

6.3.1 Case 1: 2012 Saudi Aramco Shamoon Attack (Source: Saudi Aramco Public Statement, www.saudiaramco.com)

Saudi Aramco, the largest global oil company, released a public statement acknowledging the 2012 cybersecurity incident. According to the statement, approximately 30,000 workstations were infected, but oil production and other mission-critical systems (initially protected on isolated networks) were not affected, and services were restored following a network shutdown and cleanup.

Saudi Aramco Upstream Cybersecurity strategy, calling for firewalls and well-managed network isolations of critical operations, helped avoid any operational interruptions, especially in oil and gas production and supplies.

6.3.2 Case 2: 2021 Colonial Pipeline, USA (Source: Published Reports including www.cisa.gov)

A ransomware attack led to the shutdown of the largest fuel pipeline in the U.S., highlighting how IT breaches can cascade into critical infrastructure disruption. For more information, please refer to several relevant publicly published reports.

6.3.3 Case 3: 2017 Triton Malware (Source: Public Published Reports, including www.ncsc.gov.uk)

Malware targeted a safety instrumented system, demonstrating that attacks can aim to physically damage equipment or cause safety failures.

6.3.4 Case 4: 2019 Norsk Hydro (Source: Public Published Reports including www.hydro.com)

Hydro became a victim of an extensive cyberattack in the early hours of Tuesday, March 19, 2019, impacting operations in several of the company's business areas.

These incidents are good examples, illustrating that cybersecurity is not optional but rather becoming a corporate operational imperative that must be included in any corporate risk management strategy.

6.4 DOF Risk Management

DOF Risk management is a systematic process of identifying, assessing, and controlling threats (risks) that could negatively affect the DOF infrastructure and hence derail oil companies from achieving their business objectives. As oil companies continue to expand their DOF capabilities, cybersecurity risk management must be embedded at the heart of corporate DOF risk management strategies. It is not an optional add-on but a strategic necessity that safeguards value creation, protects people and the environment, and ensures long-term operational resilience. A secure DOF is ultimately a reliable, sustainable, and trusted DOF, capable of delivering its full digital and business potential.

Cyber-resilience is as much about mindset as it is about firewalls and encryption. As the oil and gas industry moves deeper into the digital age, cybersecurity, data governance, system availability, and DR are no longer peripheral IT concerns. They are strategic corporate enablers of safe, reliable, and trusted operations. Smart and intelligent fields must be secured not just against technical failures but against deliberate attacks and data mismanagement. Operators that invest in robust cyber security and governance frameworks are not only protecting their assets, but rather they are building a foundation of trust that will support future innovation, stakeholder confidence, and regulatory compliance. By combining technology, policy, cybersecurity awareness culture, and best practices, organizations can mitigate risk and sustain continuous, efficient oil and gas operations.

References

1. National Institute of Standards and Technology (NIST). Framework for Improving Critical Infrastructure Cybersecurity. Gaithersburg (MD): NIST; 2018.
2. International Society of Automation (ISA). ISA/IEC 62443: Security for Industrial Automation and Control Systems. Research Triangle Park (NC): ISA; 2018.
3. International Organization for Standardization. ISO/IEC 27001: Information Security Management Systems. Geneva: ISO; 2022.
4. American Petroleum Institute. API Standard 1164: Pipeline SCADA Security. Washington (DC): API; 2021.
5. American Petroleum Institute. API Standard OG-86: Cybersecurity for Upstream Operations. Washington (DC): API; 2019.
6. MITRE Corporation. MITRE ATT&CK for ICS: Cyber Adversary Tactics and Techniques. McLean (VA): MITRE; 2020.
7. Dragos Inc. Industrial Control System (ICS) Threat Activity Reports. Arlington (VA): Dragos; 2019–2024.
8. Symantec Security Response. ICS and OT Cybersecurity Threat Landscape Report. Mountain View (CA): Symantec; 2020.

9 Microsoft Azure Security Center. Cloud and Hybrid Security for Critical Infrastructure. Redmond (WA): Microsoft; 2021.
10 Amazon Web Services. Security Best Practices for AWS Industrial and IoT Workloads. Seattle (WA): AWS; 2022.
11 Cisco. Securing Industrial Networks for Critical Energy Infrastructure. San Jose (CA): Cisco Systems; 2021.
12 Fortinet. OT/ICS Security Framework and Zero-trust for Industrial Environments. Sunnyvale (CA): Fortinet; 2020.
13 Palo Alto Networks. ICS/SCADA Security for Industrial Enterprises. Santa Clara (CA): Palo Alto; 2021.
14 Boyer SA. SCADA: Supervisory Control and Data Acquisition. 4th ed. Research Triangle Park (NC): ISA; 2016.
15 Stouffer K, Falco J, Scarfone K. Guide to Industrial Control Systems (ICS) Security. Gaithersburg (MD): NIST; 2015. (NIST SP 800-82).
16 Verizon. Data Breach Investigations Report (Energy and Industrial Sector). New York (NY): Verizon; 2019–2024.
17 Gartner Research. Cybersecurity for OT/ICS: Trends and Risk Management in Critical Infrastructure. Stamford (CT): Gartner; 2020.
18 World Economic Forum. Cyber Resilience in Oil and Gas: Sector Analysis. Geneva: WEF; 2020.
19 Deloitte. Cybersecurity Strategies for Digital Oilfields; 2021.
20 International Electrotechnical Commission. IEC 62443: Industrial Automation Cybersecurity. Geneva: IEC; 2018.

7

Enterprise DOF Workflows—Field Operation and Production Engineering

Harvesting Your Technological Investment

Digital oil fields (DOFs) are more than technological upgrades—they are operational ecosystems built around strategic, data-driven workflows. To harvest the full value of smart infrastructure, operators must implement automated, real-time workflows that align production, reservoir, facilities, drilling, and management teams with their corporate financial, HSE, and sustainability objectives/performance. The chapter is written based on my 35+ years of rounded experience, plus generic references to publicly published NOC and IOC articles [1–5]. In addition, this chapter will focus more on workflows that significantly transformed the way the industry manages the DOF operation and production engineering, resulting into saver operations, higher production, and outstanding HSE and sustainability performances, addressing the following strategic business layers [1, 2]:

- **Field Monitoring and Alerting Analytics Layer**: Includes AI/ML algorithms, digital twins, and multidomain technical workflows such as real time well testing, HSE dashboard, well integrity, ESP optimization, operational monitoring and alerting agents that turn data into actionable operational and engineering insights.
- **Engineering Workflows Layer**: Where day-to-day operational, production engineering, reservoir management, and facility studies and workflows are conducted, such as well performance, well testing, reservoir monitoring and real-time reservoir management, and strategic business and engineering processes.
- **Strategic Predictive and Field Optimization Layer**: Where strategic business and engineering processes (e.g., asset long-term planning, corporate safety and environmental management, finance and surface/subsurface assets, and integration and optimization studies) are conducted through real-time and static data integration.

The chapter explores the structure, objectives, and benefits of several intelligent oil field production engineering and operations workflows, with emphasis on:

- Well integrity.
- Production optimization.
- Monitoring and optimization of electric submersible pumps (ESPs).
- Virtual Multiphase Flow Meters (MPFMs).
- Water Injection Monitoring and Injectivity Analysis.

Other relevant workflows can be found in different applications based on each operator's DOF strategy [1, 4, 5].

7.1 Well Integrity Using DOF Technologies

Well integrity is a critical pillar of oil and gas operations, encompassing the application of engineering, operational, and organizational solutions to ensure that wells operate safely and efficiently throughout their lifecycle (Figure 7.1) [1, 2, 4, 6–8].

In recent years, DOF technologies have revolutionized the way companies manage well integrity, enabling real time surveillance, predictive analytics, and automated interventions. This section explores how DOF technologies are applied in well integrity management, focusing on real-time data acquisition, key performance indicators (KPIs), dashboards, and field-level integration. Specific examples from NOCs and IOCs illustrating the tangible benefits of digital transformation in this domain will be discussed.

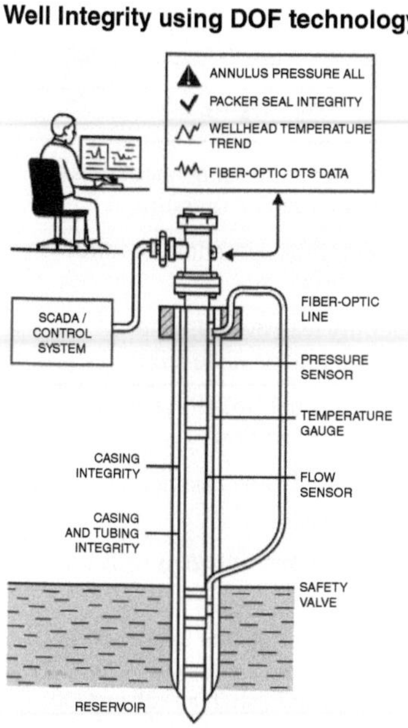

Figure 7.1 A typical DOF well monitoring and integrity data-driven model (AI generated).

7.1.1 Foundations of DOF in Well Integrity

DOF technologies integrate:

- Real-time data acquisition via downhole sensors and surface equipment.
- Advanced analytics using machine learning and predictive modeling.
- Visualization tools for situational awareness and decision-making.
- Integration of validated and QCed real-time data with other related static and well models, including the most recent calibrated digital twins.
- Workflow automation for proactive well integrity interventions.

Major national and international oil companies have developed and deployed multiple, fully integrated, real-time, well-integrity platforms, such as the advanced well integrity monitoring system (AWIMS) and intelligent field automation system, to monitor thousands of wells in real time, despite their scattered geographical and harsh remote areas.

7.1.2 Real-time Data for Well Integrity

The AWIMS approach includes gathering multidomain data across time and depth of all factors that can be related to well integrity. Examples of typical real-time data streams used for integrity monitoring are listed in Table 7.1.

Table 7.1 Typical DOF real-time data stream's integrity.

Data Type	Description	Use in Well Integrity
Tubing and casing pressure	Surface and downhole pressure readings	Monitor for pressure anomalies or sustained casing pressure
Temperature logs	Distributed temperature sensing (DTS)	Detect leaks, crossflow, or inflow anomalies
Acoustic signals	Fiber-optic distributed acoustic sensing (DAS)	Identify leak events, sand production, or valve failure
Flow rate and choke data	Real-time multiphase flow sensors	Ensure zonal isolation and detect unauthorized flow
Annular pressure monitoring	Pressure in annular spaces	Detect sustained casing pressure or gas migration
Cement bond logs (CBLs)	Digitalized cement evaluation logs	Assess cement integrity over time
Surface and subsurface valves status	Testing the integrity of all relevant surface and subsurface valves	Failure to close due to any reason is crucial for well integrity, especially critical valves such as subsurface safety valves, well tree master valves, wing valve, and others
Wellhead, wellsite emissions, leaks, and H_2S	Wellhead sensors and possibly satellite-imaging detecting emissions, hydrocarbon leaks, and H_2S in real time	Any positive detection of emissions, leaks, and H_2S releases

Table 7.2 A typical DOF real-time well integrity and surveillance KPI.

KPI	Description
Annulus pressure index	Ratio of actual to allowable annulus pressure
Sustained casing pressure frequency	Count of wells showing repeated SCP across X period
Well integrity downtime	% of time a well is shut due to integrity issues
Integrity alarm response time (mins)	Time to respond to integrity alarms
Leak detection confirmation time	Time to verify and classify a leak post-detection
Well tree master and SSSV status	Capability to close 100% in case of emergency
Wellhead, wellsite emissions, and H_2S	Capability to close 100% in case of emergency

7.1.3 KPIs for Real-time Integrity Surveillance

The AWIMS employs a series of automated integrity KPIs integrated into its DOF platforms (Table 7.2). These KPIs are continuously monitored in real time 24 × 7 and ranked by criticality.

These KPIs are visualized and tracked across well, field, and enterprise levels using real-time generated dashboards.

Other well integrity critical parameters can also be monitored with report/alert/escalate, such as:

- CO_2 emission and oil spills.
- Well integrity critical index.
- Corrosion index.
- Master valve + other valve leaks.
- H_2S and other uncontrolled gas releases, including heat maps of the affected areas based on current weather conditions.

7.1.4 Dashboards and Visualization Tools

The AWIMS leverages interactive dashboards integrated with all DOF centers, built with tools like the plant information (PI) system, database, and custom company–specific built interfaces.

7.1.5 Well-level Dashboard (Recommended Example, Table 7.3)

Similar field-level dashboard and KPIs can also be developed and implemented as part of the DOF well integrity platform.

Sample Well Integrity Alert Flow

1. Sensor detects abnormal annulus pressure, H_2S, oil spill, or gas leak.
2. Alarm triggered in the well-level dashboard (red colored).
3. KPI breach logged.

These systems enabled proactive rather than reactive maintenance, improving safety, well integrity, and reducing wells' downtime and operation costs.

Table 7.3 A typical DOF real-time well-level integrity and surveillance dashboard.

Section	Data Displayed
Real-time pressure plot	Tubing, casing, and annular pressure over time
Well integrity index or integrity status light	Options: green (OK), orange (monitor), red (immediate action required) for all wells
Alerts log	Detection, pressure imbalance, valve failure
Last integrity test	Summary of pressure test, casing leaks, or mechanical integrity test
Escalation capability	Escalation to a higher management level of any critical well that was not attended within the specific time frame
Integrity heatmap	Well-by-well status: criticality coded by color
KPIs summary panel	Sustained casing pressure (SCP) frequency, downtime %, leak confirmations
Historical trends	Monthly stats for field-wide integrity KPIs
Action tracking	Status of pending and completed remedial actions

Each operator can also implement predictive integrity models based on machine learning trained on historical integrity failure data:

- **Failure Mode Prediction**: Identify wells likely to develop corrosion.
- **Automated Ticketing**: Auto-generation of maintenance tickets based on alarm validation.
- **Digital Twin Models**: Simulate well integrity under different stress and failure scenarios.

These systems are integrated into the operator's operational command centers, providing centralized oversight and rapid response capabilities. Furthermore, each operator might have its own developed risk calculation factor using its own developed formula of well integrity index (0–100%). The risk factor could be a reactive approach using the current real-time data, where it is calculated every five seconds, or a proactive approach using additional predictive models (e.g., AI, ML, or advanced analytics). Operators may also classify different categories of the risk factor with a critical threshold % (e.g., normal, low risk, high risk). Figure 7.2 shows a typical predicted well integrity risk factor with 70% risk factor threshold of critical status.

This chart shows the predicted well integrity risk scores over the next seven days for five wells:

- The red dashed line at 70% marks the high-risk threshold.
- Wells approaching or exceeding this line are flagged for proactive inspection or intervention with alert and time escalation algorithms/smart agents.

This kind of predictive output, used in platforms like this, helps engineers plan ahead and avoid unplanned shutdowns or failures. DOF technologies have significantly enhanced well integrity management, shifting from periodic manual testing to continuous, real-time surveillance. NOCs and IOCs are pioneering the use of

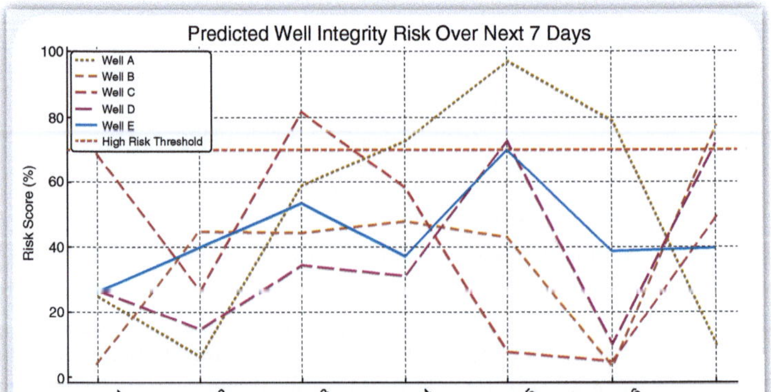

Figure 7.2 Typical example of a well integrity visual chart.

real-time data, advanced KPIs, and integrated dashboards, setting a benchmark for the industry. By coupling digital monitoring with predictive analytics, companies can ensure safer operations, lower intervention costs, and better asset longevity.

7.2 Well Production Optimization (ICVs) and DOF Technologies

Long horizontal wells with several reservoir compartments and multilateral wells—where multiple laterals branch from a single main bore—allow operators to access larger reservoir volumes with fewer surface penetrations, thereby increasing productivity, reducing costs, and reducing environmental footprint. However, managing production from these laterals effectively poses significant challenges, especially with regard to inflow balancing, water or gas breakthrough, and reservoir sweep efficiency that could impact the well productivity and overall reservoir recovery factor [1, 2, 9–14].

To address these challenges, inflow control valves (ICVs) are installed in each lateral, providing a mechanism to dynamically control or shut off flow from specific lateral or/an zones within each lateral. When integrated with DOF technologies and real-time surveillance systems, ICVs enable continuous, automated optimization of production (Figure 7.3).

DOF and real-time data, when integrated with calibrated well models (digital well/reservoir twins), enable engineers to automatically evaluate the well/completion performance of existing oil producers and water injectors completed with this special equipment (ICVs, ICDs, etc.) and help to increase oil production, decrease undesired water and gas production. This needs well-calibrated digital twin models in addition to the real-time QCed DOF data.

7.2.1 Understanding ICVs in Multilateral Wells

ICVs are downhole devices installed along the completion string of each lateral. They can be operated:

- Manually (via well intervention).
- Remotely (via hydraulic/electric actuation).
- Automatically (smart ICVs with control algorithms).

The use of smart ICVs—combined with real-time downhole sensing and surface control systems—has become a key enabler in managing production from multilateral wells.

7.2.2 Integration of ICVs with DOF Systems

Table 7.4 shows a typical integration of components and functionality of ICVs within a DOF environment.

Figure 7.3 A typical DOF production optimization in multilateral wells (AI generated).

Table 7.4 A typical ICV and DOF integration components and functionality.

Component	Functionality
Downhole sensors	Measure pressure, temperature, flow rate, and water cut in real-time
Real-time data transmission	Fiber-optic systems relay data to the surface
Centralized DOF platform	Collects, analyzes, and visualizes production data from each lateral
Optimization algorithms	Suggest or execute valve actuation for maximum oil and minimal water/gas

Table 7.5 A typical DOF sensor data used in a typical ICV optimization workflow.

Sensor Data	Use Case
Zonal pressure differential	Indicates potential for crossflow or water encroachment
Zonal flow rate	Confirms actual production per lateral
Water cut/gas–oil ratio (GOR)	Triggers inflow restriction to reduce W&G production
Downhole temperature	Helps detect fluid movement or gas breakthrough

Major oil companies integrate ICVs with their DOF platform, enabling centralized control of thousands of wells, maximizing the added value of multilateral well configurations.

7.2.3 Real-time Data Streams for ICV Management

Table 7.5 shows a typical DOF key parameter captured in real time to support production engineering and/or reservoir management ICV optimization workflow.

DOF systems visualize this data through production dashboards and allow remote valve adjustments via supervisory control systems.

7.2.4 AI/ML-based Optimization of ICV Settings

Real-time data are fed into AI/ML models that learn the production dynamics of each lateral and recommend optimal valve positions.

AI applications include:

- **Water Breakthrough Prediction**: Shut or choke back the lateral before water dominates.
- **Zonal Balancing Models**: Ensure uniform drawdown to delay gas coning.
- **Production Maximization Algorithms**: Adjust ICVs to maintain plateau production.
- **Autonomous Valve Control**: Fully automate valve changes based on real-time thresholds.
- Benefits include:
  - Delay water breakthrough.
  - Increase oil recovery.
  - Reduce well intervention cost.

7.2.5 ICV Optimization Dashboard Example

A typical ICV optimization dashboard in a DOF interface would show (Table 7.6).

Figure 7.4 shows a typical multilateral well with four laterals, each of which is controlled by one ICV, where extensive ICV optimization tests are conducted. Each plot shows each lateral status (open, closed), ICV setting, water cut, and pressure drop across the ICV. It shows an optimized setting where lateral 2 is completely closed due

Table 7.6 A typical DOF ICV optimization dashboard.

Panel	Displayed Info
Lateral overview map	Real-time flow status, open/closed valve indicators
Zonal performance chart	Flow rate versus pressure drawdown for each lateral
Anomaly detection panel	Alerts for sudden flow drops, water spikes, or crossflow
ICV control interface	Manual or AI-driven valve control options
Optimization KPIs	Zonal oil rate, water cut %, ICV setting %, net oil gain

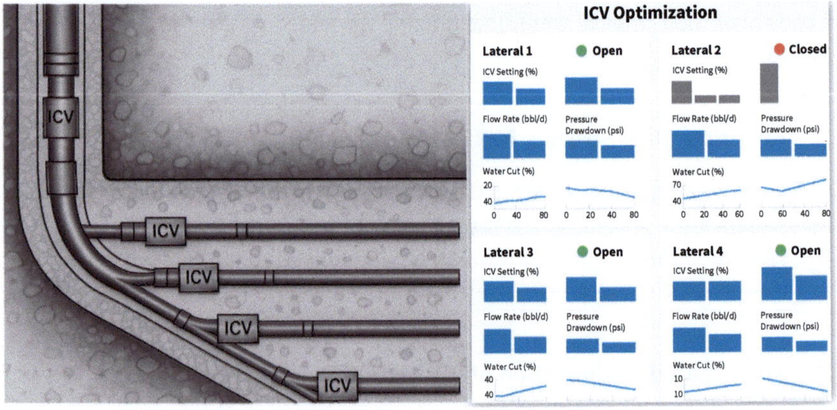

Figure 7.4 Typical example of a generic ICV-optimized multilateral well with four laterals.

to the very high water cut. These dashboards allow engineers to remotely optimize individual laterals in near real time to increase recovery and reduce development cost.

The following is the step-by-step autonomous ICV adjustment workflow:

1. Water cut spike detected in lateral 2 via real-time sensor.
2. **DOF AI Model Evaluates**: Compares to the threshold and the reservoir model forecast.
3. **ICV Adjustment Recommended**: Choke lateral 2 by 50% to reduce water inflow.
4. **Remote Command Executed**: ICV adjusted from the surface SCADA system.
5. **Performance Monitored**: Net oil gain and reduced water cut confirmed within a few hours-days to reach sustainable production, representing steady state flow based on the well reservoir permeability (the higher the permeability is, the shorter the well reaching state).

7.2.6 Reservoir Management Benefits

The dynamic DOF ICV/ICD optimization workflow supports long-term reservoir health by:

- Preventing uneven depletion across layers.
- Managing reservoir sweep through intelligent injection/production control.

- Minimizing water coning or gas channeling.
- Avoiding excessive drawdown in high-permeability streaks.
- Avoiding any workover due to high water cut or gas–oil ratio.

More benefits with much higher production/recovery and HSE performance can be achieved if the following emerging trends in multilateral well optimization are added:

- Self-learning completions with embedded edge AI.
- Digital twins for real-time simulation of fluid front progression.
- Self-adjusted ICVs with ultra-precise zonal isolation capabilities.
- Closed-loop optimization combining reservoir simulators, surface facilities simulators, and live ICV control.

Multilateral wells offer significant advantages in reservoir contact and production potential. However, their effectiveness depends on precise and dynamic control of each lateral. Real-time ICV optimization, combined with AI-driven analytics, helps companies to further:

- Increase production and recovery factors.
- Delay reservoir damage and undesired fluids (minimize water/gas production).
- Improve overall well performance and economics.

7.3 Monitoring and Optimizing Electric Submersible Pumps

Electric submersible pumps (ESPs) are a widely used form of artificial lift in oil wells, especially in high-volume and offshore operations. Their efficient operation is critical to maintaining production targets and managing lifting costs. However, ESPs are prone to failures, trips, and inefficiencies due to complex well dynamics and operational stresses.

With the advent of DOF technologies and artificial intelligence/machine learning (AI/ML), operators can now monitor, predict, and optimize ESP performance in real time. Major NOCs and IOCs, with the help of ESP manufacturers and IT service providers, are leading the way in deploying these advanced capabilities [1, 2, 15–21].

7.3.1 Smart ESP Monitoring in DOFs

Modern ESP monitoring systems integrate multiple data streams to provide real-time insights into equipment health and production behavior (Figure 7.5).

Key parameters include:

- ESP intake pressure.
- ESP discharge pressure.
- Flowing wellhead pressure (FWHP).
- Pump frequency (Hz).
- ESP pump efficiency.
- Motor current and voltage.
- Other pump operational parameters.

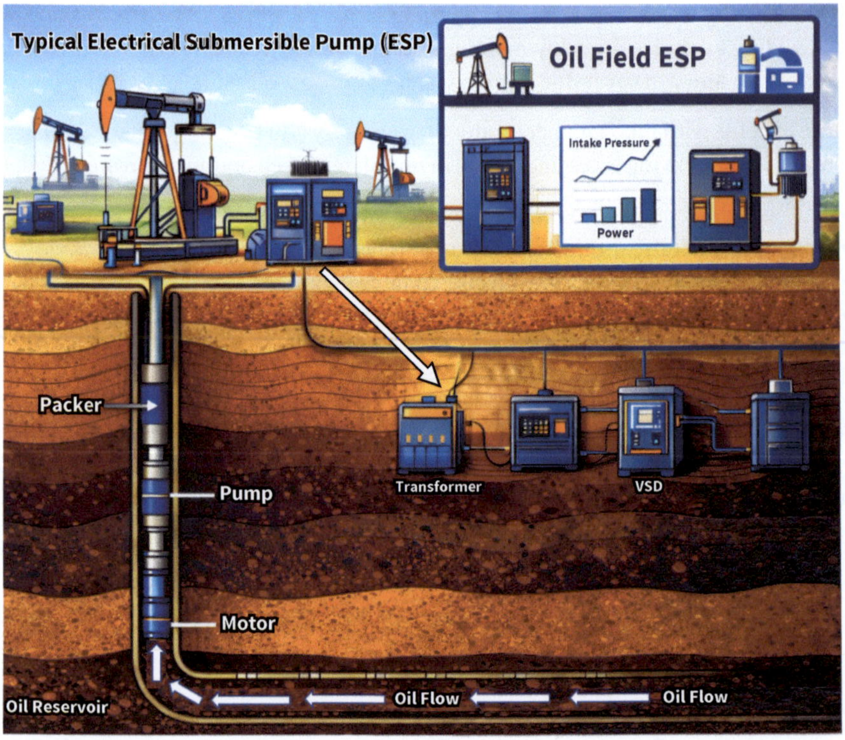

Figure 7.5 A typical DOF ESP components and optimization diagram (AI generated).

7.3.2 Components of ESP Failure Monitoring

- Trip detection (automatic shutoffs due to abnormal conditions).
- Motor load trend analysis.
- Temperature and vibration surveillance.
- Power consumption anomalies.

These measurements are continuously logged via SCADA systems and analyzed for deviations, failures, or optimization opportunities.

7.3.3 Components of ESP Digital Optimization

ESP optimization using DOF and AI involves:

- Real-time data acquisition.
- Predictive maintenance and failure modeling.
- Pump trip prevention.
- Power efficiency optimization.
- AI-driven advisory systems and automation.

7.3.4 Advantages of ESP Optimization Workflows (Table 7.7)

Table 7.7 Typical advantages of classical DOF ESP optimization workflow.

Benefit	Impact
Increased production	Better drawdown, optimized reservoir contact
Reduced energy cost	Align operation and avoid overpumping
Enhanced equipment life	Reduced motor stress and heat generation
Early warning on potential failures	Downtime prevention through predictive analytics
Centralized decision support	Enables field-wide optimization from control centers

Table 7.8 Typical DOF ESP real-time data acquisition parameters.

Parameter	Description
Intake and discharge pressure	Indicates pump loading and fluid flow conditions
Pump motor current and voltage	Identifies power consumption and anomalies
Pump speed (Hz)	Tied to production rate and variable speed drives
Downhole temperature	Detects overheating and possible gas locking
Vibration and harmonics data	Tracks mechanical degradation
Fluid rate and water cut	Used for performance modeling

7.3.5 Real-time Data Acquisition in DOF

Modern ESPs are equipped with downhole sensors and integrated with surface SCADA systems to provide continuous data feeds (Table 7.8).

ESP data are transmitted in real time, where pump curves are dynamically generated, and pump run efficiency can be scientifically quantified to enforce high efficiency operation, which can result in longer runtime and higher rate. Figure 7.6 shows a typical ESP chart showing its operational conditions as per the pump manufacturer's operating curve, comparing it with the optimum pump efficiency.

Major oil companies with DOF, for example, integrate this data into their advanced ESP monitoring system (AEMS) under its broader DOF framework.

7.3.6 ESP Failure Prediction Using AI/ML

AI and ML models have been developed and trained by several ESP manufacturers and/or major operators using years of historical ESP run and failure historical data to predict likely failure events such as:

- Bearing wear.
- Seal failure.

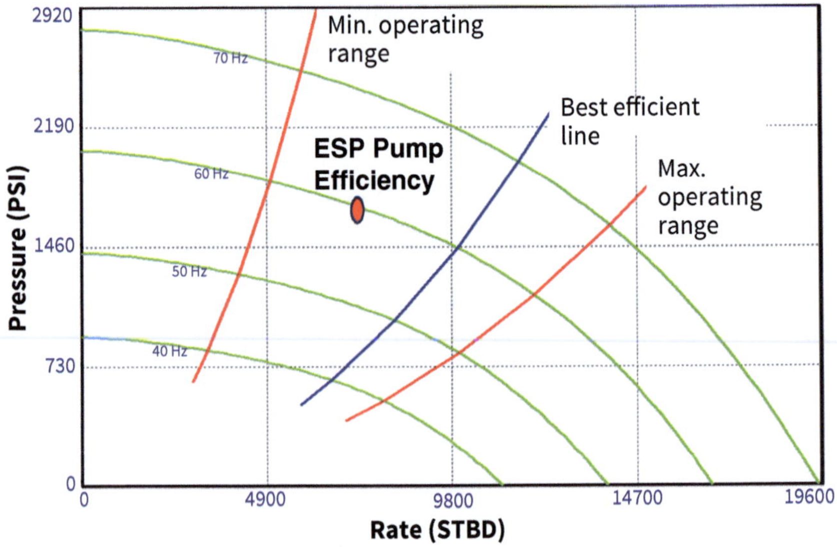

Figure 7.6 Typical ESP pump operation and pump efficiency curve.

- Cable insulation breakdown.
- Gas locking.
- Scale or sand accumulation.
- Other mechanical, operational, and/or electrical failure reasons.

7.3.7 ML Algorithms Commonly Used

The following are some typical standard ML algorithms that can be used for ESP optimization:

- **Random Forest**: Classifying failure modes.
- **Gradient Boosting**: Early warning signals for degradation.
- **Autoencoders**: Detecting anomalies in multidimensional data.

Case Study (Aslam & Reddicharla, 2023; *Journal of Petroleum Technology*, 2024)

ADNOC has developed and deployed AI/ML models for ESP failure prediction built from multiple years of sensor data and is now rolling out across hundreds of thousands of wells. These models have achieved failure prediction accuracies between 75% and 87%, depending on the well type and operating conditions. Importantly, they provide lead times ranging from two weeks to three months before actual pump failure, enabling proactive interventions such as parameter optimization or timely pull-and-replace operations.

The practical benefits of these predictive systems are significant. ADNOC field pilots have demonstrated a 20–25% reduction in unplanned downtime and an increase in well availability of approximately 10%. When deployed at scale, these systems will contribute substantially to ADNOC's broader digital transformation agenda. Indeed, ADNOC credited AI-powered predictive maintenance tools, including ESP failure prediction, with unlocking hundreds of millions of dollars in operational value in 2023.

7.3.8 Pump Trip Prediction and Prevention

Pump trips often occur due to:

- Voltage fluctuations.
- Sudden load changes.
- High gas–liquid ratio (GLR).
- Excessive vibration.
- Overcurrent from sand ingress.
- Other mechanical, operational, and/or electrical failure reasons.

7.3.9 AI-based Trip Prevention

ML models, particularly time-series forecasting with recurrent neural networks, are used to predict imminent trips based on patterns in:

- Power consumption spikes.
- Transient pressure oscillations.
- Discharge temperature rise.
- Subtle increases in vibration amplitude.

7.3.10 ESP Power Optimization

Power consumption in ESP operations is a significant operational cost. AI/ML and DOF systems optimize power use by:

- Adjusting variable speed drives (VSD) dynamically.
- Balancing production rate versus power draw.
- Identifying gas interference or suboptimal frequency ranges.

7.3.11 Dashboards and Visualization

Several NOC and IOC companies integrate real-time ESP data and AI insights into intuitive dashboards:

7.3.12 ESP Performance Dashboard Elements

These dashboards (Table 7.9) are used not only for monitoring but also for remote control and automation of ESP systems.

Table 7.9 Typical elements of an ESP performance dashboard.

Module	Content
Real-time pump metrics	Pressure, speed, temperature, and voltage
AI advisory panel	Failure risk, trip prediction, and power efficiency
Historical performance trends	Run life, previous failures, and maintenance logs
Optimization recommendations	VSD setting changes and cleaning alerts

7.3.13 Workflow Integration and Remote Operations

Integrated workflows allow AI-based alerts to automatically:

- Trigger diagnostic workflows.
- Schedule pull/replacement.
- Adjust VSDs remotely.
- Notify engineers via mobile apps or control centers.

Remote ESP operations centers monitor thousands of wells across multiple fields, enabling fully digital lift system management.

The next frontier in ESP optimization includes:

- **Digital Twins of ESPs**: Simulate real-time physical and chemical conditions.
- **Reinforcement Learning**: For adaptive frequency and production tuning.
- **Edge AI Deployment**: Localized models running near the wellsite.

DOF and AI/ML technologies are transforming ESP operations from reactive to predictive and optimized. Real-time monitoring, AI-based failure prediction, trip forecasting, and energy efficiency are now industry standards among leading operators. As fields move toward autonomous operations, ESP monitoring and optimization represent a high-impact area, where digital technologies yield measurable production and economic benefits.

IOCs and NOCs are demonstrating how intelligent ESP management leads to increased uptime, reduced cost, and enhanced safety, establishing a new digital standard in artificial lift operations.

7.4 Virtual MPFMs in DOFs

In the DOF era, real-time production monitoring is crucial for optimizing operations and ensuring accurate reservoir management. While physical MPFMs are ideal for real-time flow measurements, their high cost, maintenance complexity, and limited scalability across large fields present significant constraints (Figure 7.7) [1, 2, 5, 9, 20, 22–28].

As a cost-effective alternative, virtual flow metering (VFM) systems have emerged—providing continuous estimation of oil, gas, and water production using

Figure 7.7 A typical oil field MPFM (AI generated).

validated real-time data, as part of the DOF, integrated with advanced mathematical and digital models.

7.4.1 What Is VFM?

VFM involves using real-time surface and downhole data—such as pressure, temperature, and choke size—combined with mathematical models or digital twins to estimate three-phase flow rates (oil, water, gas) without installing physical meters.

Key DOF Inputs for VFM

- Tubing head pressure (THP).
- Bottom-hole pressure (BHP).
- Choke setting.
- Water cut (WC) and GOR.
- Real-time sensor data (PDHMS, SCADA).
- Well productivity index and calibrated digital twin (in certain models).

7.4.2 Benefits and Limitations of VFM

Benefits

- Reduces the cost of MPFM deployment and maintenance.
- Enables real-time monitoring across entire fields.
- Accelerates production allocation.
- Provides rapid response to operational changes.
- Accuracy within ~90% if models are well-calibrated.

Limitations

- Dependent on the quality of input data and model calibration.
- Requires periodic updates for changing reservoir or well conditions.
- Less reliable than MPFMs in highly transient or unstable wells.

7.4.3 VFM Modeling Approaches

Model 1: Choke Models (Gilbert-type Equations)

- Empirical correlations (e.g., Gilbert, Ros, Perkins) estimate single-phase flow rate through a choke.
- Key parameters: choke size (in 64ths), THP, and GOR.
- General form:

$$\frac{1}{(Q_L)\mathrm{Surf}} = \frac{a_1 \times \mathrm{THP} \times \mathrm{GOR}^{a_2}}{\left(\frac{d}{64}\right)^{a_3}} \tag{7.1}$$

- Where Q_L-Surf is the total liquid rate at surface condition in STBD; THP, the tubing head pressure in psi; GOR in SCF/STB, d, the current choke size over 64 in.; a_1, a_2, and a_3, the correlation-specific flow coefficients given in Table 7.10.

Table 7.10 Parameters (a_1, a_2, a_3) for several standard surface choke correlations.

Correlation	a_1	a_2	a_3
Gilbert	10.00	0.546	1.89
Rose	17.4	0.500	2.00
Baxendell	9.56	0.546	1.93
Achong	3.82	0.650	1.88

Model 2: Wellbore Digital Twin

- Simulates vertical lift performance (VLP) from reservoir perforations to the surface.
- Accounts for:
  - Well trajectory (MD, TVD).
  - Tubing diameter and roughness.
  - Geothermal gradients.
  - WC and GOR.
- Models pressure drops using:

$$\frac{d_p}{d_L} = \frac{g}{32.2}\rho \ \text{Sin}\theta + \frac{\rho v}{32.2}\frac{d_v}{d_L} + f\frac{\rho v^2}{2g_c d} = \text{Gravity} + \text{Expansion} + \text{Friction} \quad (7.2)$$

where dp/dL is the total pressure losses from a node at a perforation to a node at the surface in psi/ft.; g the gravitational acceleration in ft./s^2; ρ the density of the mixed fluid in lbm/ft^3; θ the wellbore angle; v the fluid velocity across the tubing in ft./s; f the Moody friction factor (dimensionless); d the internal tubing diameter in inches; and dv/dL is the velocity differential per length.

- This process generates pressure versus rates tables or curves (VLP) for multiple wellhead pressures to be used to predict the rate on a real-time basis based on the current DOF real-time QCed wellhead pressure.
- Tables need to be recalculated when well conditions, such as water cut change (e.g., monthly, quarterly, annually)

Model 3: Polynomial *P*–*Q* Curve

- Uses second-order polynomial to fit pressure-rate (*P*–*Q*) data: $Q = a_0 + a_1 P + a_2 P^2$, where a_0, a_1, and a_2 are coefficients determined by matrix regression using real-time *P*–*Q* data pairs and can be calculated as follows:

$$\begin{bmatrix} a_0 \\ a_1 \\ a_2 \end{bmatrix} = inv \begin{bmatrix} n & \Sigma P_i & \Sigma P_i^2 \\ \Sigma P & \Sigma P_i^2 & \Sigma P_i^3 \\ \Sigma P_i^2 & \Sigma P_i^3 & \Sigma P_i^4 \end{bmatrix} * \begin{bmatrix} \Sigma Q_i \\ \Sigma Q_i P_i \\ \Sigma Q_i P_i^2 \end{bmatrix} \quad (7.3)$$

where n is the number of PQ pairs

- Quick to apply but needs several rate tests and frequent recalibration.

Example: Construct a curve-fitting polynomial using the following *PQ* dataset.

Q	P
40	880
2100	820
4200	740
6340	630
8445	503
10550	370
12650	230
14750	90

	P	P^2	P^3	P^4		Q	PQ	QP^2
1	880	774400	681472000	5.997E+11		40	35200	30976000
2	820	672400	551368000	4.5212E+11		2100	1722000	1.412E+09
3	740	547600	405224000	2.9987E+11		4200	3108000	2.3E+09
4	630	396900	250047000	1.5753E+11		6340	3994200	2.516E+09
5	503	253009	127263527	6.4014E+10		8445	4247835	2.137E+09
6	370	136900	50653000	1.8742E+10		10550	3903500	1.444E+09
7	230	52900	12167000	2798410000		12650	2909500	669185000
8	90	8100	729000	65610000		14750	1327500	119475000
Σ	4263	2842209	2.079E+09	1.5948E+12		59075	21247735	1.063E+10

$$\begin{bmatrix} a_0 \\ a_1 \\ a_2 \end{bmatrix} = inv \begin{bmatrix} 8 & 4263 & 2842209 \\ 4263 & 2842209 & 2078923527 \\ 2842209 & 2078923527 & 1.5948E+12 \end{bmatrix} * \begin{bmatrix} 59075 \\ 21247735 \\ 1.063E+10 \end{bmatrix}$$

$$\begin{bmatrix} a_0 \\ a_1 \\ a_2 \end{bmatrix} \begin{bmatrix} 1.724495 & -0.00727689 & 6.412420E-06 \\ -0.00727689 & 6.412420E-06 & -3.69150E-08 \\ 6.412420E-06 & -3.69150E-08 & 3.73193E-11 \end{bmatrix} * \begin{bmatrix} 59075 \\ 21247735 \\ 1.063E+10 \end{bmatrix} \rightarrow$$

$$\begin{bmatrix} a_0 \\ a_1 \\ a_2 \end{bmatrix} \begin{bmatrix} 15413.995 \\ -9.1458 \\ -0.00888 \end{bmatrix} \rightarrow Q = 15413.995 - 9.1458\, P - 0.00888\, P^2$$

$$Q = 15413.995 - 9.1458\, P - 0.00888\, P^2$$

The second-degree polynomial can be plotted as shown in Figure 7.8.

- Using the real-time flowing wellhead pressure (FHP), you can predict the flow rate at any time.
- The polynomial needs to be reconstructed periodically.

Model 4: Near-wellbore Inflow Performance Relationship

- The most common method for modeling this zone is the use of the inflow performance relationship (IPR), which calculates the flow rates at bottom

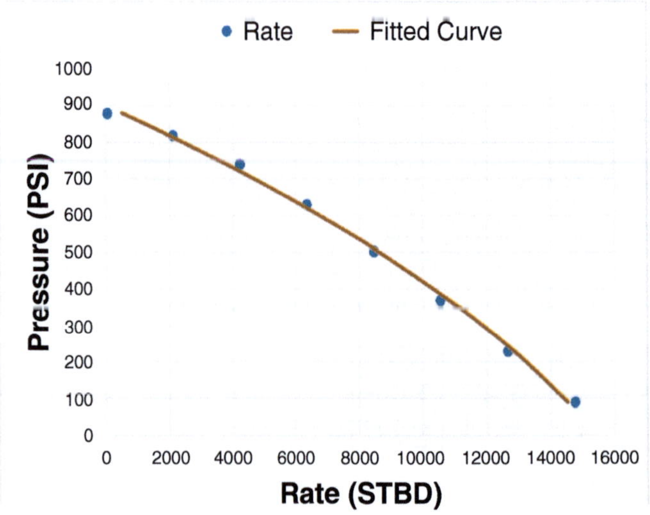

Figure 7.8 A typical *P/Q* curve.

hole (corrected PDHMS to mid-perforation), assuming a constant reservoir-pressure-boundary condition (steady state or pseudo-steady state). The expression estimates the ability of the reservoir to deliver fluids to the wellbore using the famous straight-line IPR:

$$\mathrm{PI} = \frac{Q_{\mathrm{tot}}}{P_e - f_{\mathrm{BHP}}} \tag{7.4}$$

where:

- Q_{tot}: Total rate in STB/D
- PI: Productivity index in STB/D/psi
- P_e: Reservoir pressure in psi
- f_{BHP}: Flowing bottom-hole pressure in psi

Model 5: Integrated IPR and VLP Curves

- The IPR and OPR/VLP curves can be integrated into a single plot showing f_{BHP} as a function of flow rates.
- The intersection of the curves indicates the operating point of the current well production in terms of total liquid rate, gas rate, and THP.
- The IPR/VLP models are calibrated using flow tests at separator conditions approximately every month or quarter. The below inflow performance (IPR) and outflow performance (OPR) curves can be automatically generated by the workflow every month for all operating wells, showing the optimum target rates for each well (Figure 7.9: The rate at the intersection between IPR and OPR curves).
- Intersection of IPR and VLP curves identifies the well's ideal operating rate.
- Requires periodic flow tests to calibrate both models.
- With this known production value set point and THP, the f_{BHP} is estimated to use the VLP model and extrapolate the downhole pressure at the perforations.
- To meet at an operating point, the VLP model must match the IPR model. For daily operations and in the absence of well tests, the choke model is used to estimate the flow rate at a given GOR, WC, and THP.
- The IPR model computes the f_{BHP} based on the rate estimated in the choke model (assuming constant reservoir pressure, WC, and GOR from the latest well test available).

Model 6: IPR + VLP + Choke Integration

- The plot shows in the top the THP as a function of fluid rate, showing both the current operating point of the wells and the value of the test.
- In the bottom (Figure 7.10), it shows the extrapolated flowing BHP as a function of fluid rates, showing VLPs using different choke sizes.
- Used when well test data is unavailable.
- Choke model estimates rate → IPR estimates BHP → VLP extrapolates pressure profile.

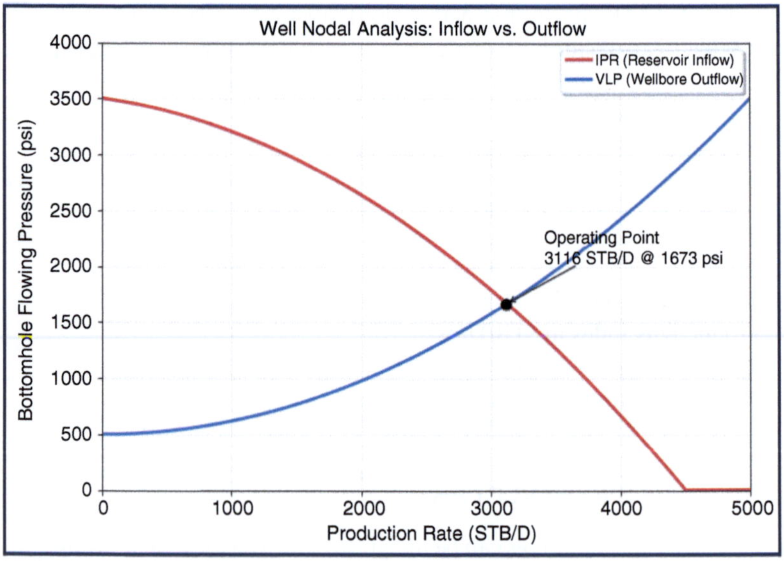

Figure 7.9 A typical IPR and OPR curves for oil-operating wells.

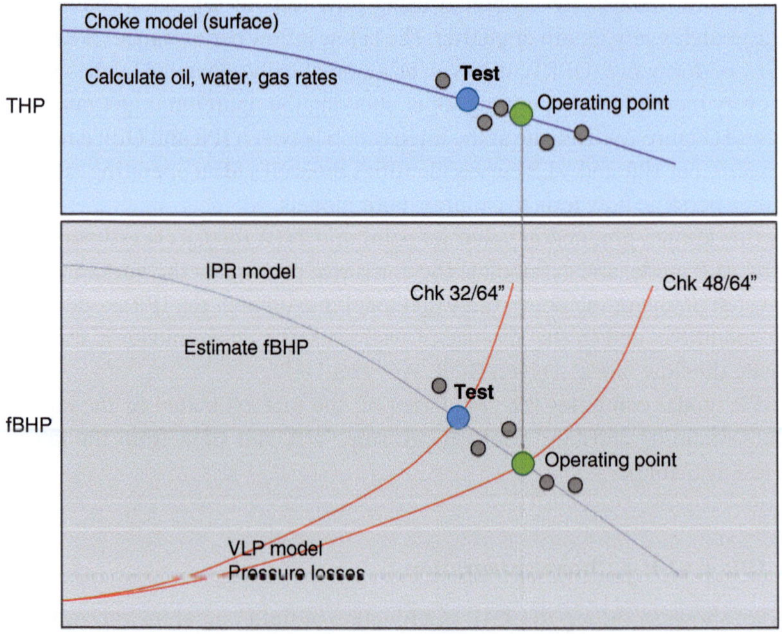

Figure 7.10 A typical choke rate model of IPR and OPR/VLP integration and optimization.

Model 7: Surface + Subsurface Network Modeling

- Field Surface and Subsurface networks can be integrated and linked to DOF data to perform full-field studies, optimizing the performance of the entire asset from reservoir to central processing facility using real-time monitoring and the capabilities of well and facility network modeling based on virtual rates calculations if physical MPFMs are not available (Figure 7.11).
- Links all wells and surface facilities in a digital twin of the entire field.
- Enables system-level optimization based on virtual rates.
- Critical for integrated asset modeling in DOF environments.
- Requires special advanced engineering skills and sometimes high-performance computing facilities
- Generates several optimization studies that could significantly optimize production and surface facilities. Furthermore, surface facilities bottlenecks can be identified and then removed.

Model 8: AI and ML

- AI is the application of several advanced ML techniques for a specific case of multiphase production rates measurements and monitoring [29, 30].

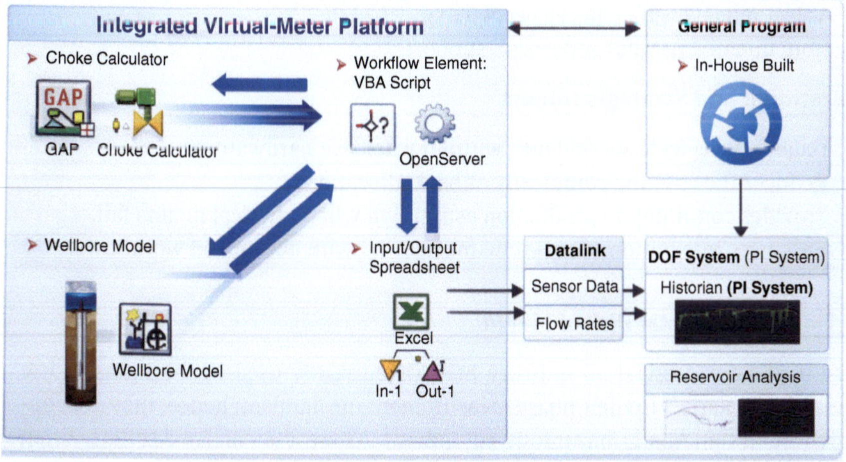

Figure 7.11 A typical DOF surface and subsurface integrated network virtual MPFM workflow (AI regenerated).

- It provides predictive capabilities of different types of ML instruments based on data analytics and ML algorithms.
- The effect of real-time data noise and data quality on the quality of estimates could be major. Hence, implementing a comprehensive and effective DOF real-time data quality system is crucial to the success of this model.
- These data-driven methods are very capable of predicting multiphase flow rates with sufficient accuracy and can be considered as a backup solution for a conventional multiphase meter.
- AI models use ML to predict flow rates using historical and real-time data.
- Algorithms include:
  - Neural networks.
  - Decision trees.
  - Regression analysis.
- Suitable as a backup to physics-based models, particularly in data-rich environments.

7.4.4 Self-maintaining VFM Systems

To maintain reliability, VFM models must be automatically updated in response to operational changes and improve their accuracy %.

Daily VFM Workflow

1. Collect 24h data (pressure, temperature, rates).
2. Clean and filter abnormal values.
3. Compute average conditions.
4. Run VFM model (choke, polynomial, digital twin, etc.).
5. Compare with historical/test data.
6. Adjust model coefficients as needed (e.g., PI, skin).
7. Store updated results for decision support.

Operational and Strategic Impact

- Enables field-wide surveillance without extensive hardware.
- Reduces the cost and complexity of back allocation.
- Provides continuity in production estimation when physical meters fail.
- Enhances integration of real-time monitoring with automated workflows.

7.4.5 Environmental Risk of MPFM

Some of the commercial oil industry MPFMs currently used in the industry use radioactive materials in their phase measurement mechanism; hence, they are considered environmentally hazardous equipment. These radioactive MPFMs, often used in oil and gas fields to measure the proportions of oil, gas, and water in production streams, utilize gamma-ray sources such as cesium-137 or cobalt-60 for phase differentiation. While these instruments provide high-accuracy data for real-time production optimization, they pose significant radiation safety risks if not properly managed. The presence of sealed radioactive sources requires strict compliance with IAEA and national radiation protection regulations, controlled handling, and regular leak testing. Accidental damage, improper shielding, or unmonitored storage can lead to radiation exposure hazards for field personnel, contamination risks, and environmental liability. Consequently, operators must ensure continuous radiation monitoring, personnel dosimetry, and secure source containment, along with trained and certified radiation safety officers to mitigate potential exposure [5].

Finally, virtual multiphase flow metering is a powerful, flexible solution in modern DOF. Through a combination of physics-based modeling and data-driven AI, VFM systems can offer real-time flow rate estimation with reasonable accuracy, enabling better-informed decisions and optimized asset management across smart fields.

7.5 Water Injection Monitoring and Injectivity Analysis in DOFs

Water injection is a fundamental part of pressure maintenance and enhanced oil recovery strategies in reservoir management. In DOFs, real-time data acquisition and automated workflows provide an opportunity to continuously monitor injectivity, optimize water allocation, and reduce uncertainty in field performance. This section focuses on smart workflows for injection well monitoring, injectivity index (II) calculations, and simplified modeling techniques using real-time field data [1, 2, 31–34].

7.5.1 Components of a Real-time Water Injection Monitoring System

To implement a smart injection monitoring workflow in a DOF, the following infrastructure is essential:

- **Wellhead Pressure Gauges**: To track real-time injection pressure.
- **SCADA Connectivity**: Ensures seamless data transmission from field to control centers.
- **Digital Well Models**: Calibrated models, pressure, and flow behavior downhole.
- **Solar Flowmeters** (Figure 7.12): For accurate surface injection rate measurement.

7.5.2 Conventional Water Injection Modeling

Modeling a water injection well is often more straightforward than production wells due to its single-phase flow nature. Key parameters required include:

- Injection rate and pressure (measured at surface).
- Reservoir properties (especially II).
- Bottom-hole pressure (SBHP).
- Tubing diameter and roughness.
- Fluid properties (density, viscosity).
- Geothermal gradient and deviation angle.

Figure 7.12 A typical water injection solar flow meter (AI generated).

Commercial simulation tools are used for detailed pressure profile modeling across the injection column.

7.5.3 Simplified DOF Real-time Modeling Approach

To enable fast, real-time monitoring without full simulation software, a simplified empirical modeling technique can be applied. We usually have the following installed in any water injection well connected to DOF: solar flow meter and choke with pressure sensors, where you can change the choke setting to reflect different injection rates (Figure 7.13) with its associated data such as injection rate and injection pressure.

7.5.4 Basic Injectivity Equation

$$Q_{\text{inj}} = \text{II} * (P_{\text{FB}} - P_{\text{SB}}) \tag{7.5}$$

where

- Q_{inj}: Surface injection rate (barrel/d) read from surface rate sensor.
- II: Injectivity index (barrel/d/psi).
- P_{FB}: Flowing bottom-hole injection pressure (psi).
- P_{SB}: Static bottom-hole pressure (psi) or reservoir pressure.

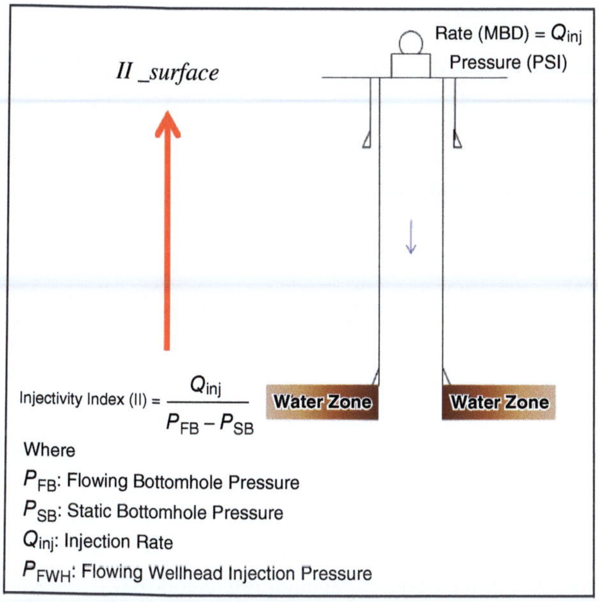

Figure 7.13 Typical water injection.

7.5.5 Surface Pressure Projection Equations

To relate surface wellhead pressure (P_{Swh}) to static bottom-hole pressure PSB, we account for:

$$\text{II} = \frac{Q_{\text{inj}}}{P_{\text{FB}} - P_{\text{SB}}} \rightarrow P_{\text{FB}} - P_{\text{SB}} = \frac{Q_{\text{inj}}}{\text{II}} \tag{7.6}$$

where $P_{\text{FB}} = (P_{\text{Fwh}} + P_{\text{Grav}} - P_{\text{fric}})$
and $P_{\text{SB}} = (P_{\text{Swh}} + P_{\text{Grav}})$

Substitute: $(P_{\text{Fwh}} + P_{\text{Grav}} - P_{\text{fric}}) - (P_{\text{Swh}} + P_{\text{Grav}}) = \dfrac{Q_{\text{inj}}}{\text{II}}$

$$\text{Simplifying: } (P_{\text{Fwh}} - P_{\text{fric}} - P_{\text{Swh}}) = \frac{Q_{\text{inj}}}{\text{II}} \tag{7.7}$$

where

- P_{Fwh}: Flowing wellhead pressure (psi) read from wellhead pressure sensor.
- P_{fric}: Friction losses (psi).
- P_{grav}: Hydrostatic or gravity head (psi).
- P_{Swh}: Shut-in wellhead pressure (psi).

Rearranging Equation 7.7:

- $P_{\text{Fwh}} = P_{\text{fric}} + \dfrac{Q_{\text{inj}}}{\text{II}} + P_{\text{Swh}}$ (7.8)
- $P_{\text{fric}} = \dfrac{0.000011466 * f * L}{d^5} Q_{\text{inj}}^2$ (7.9)

- Equation 7.7 becomes:
- $$P_{\text{Fwh}} = \frac{0.000011466 * f * L}{d^5} Q_{\text{inj}}^2 + \frac{1}{\text{II}} Q_{\text{inj}} + P_{\text{Swh}} \tag{7.10}$$

where

- f: Friction factor (depends on flow regime).
- L: Tubing length (ft).
- d: Tubing diameter (inch).

This equation enables a quick calculation of pressure losses along the wellbore. Further simplification of Equation 7.10, where:

$$A = \frac{0.000011466 * f * L}{d^5}$$

$$B = \frac{1}{\text{II}}$$

$$C = P_{\text{Swh}}$$

The outcome equation is a second-degree polynomial equation as follows:

$$P_{\text{Fwh}} = AQ_{\text{inj}}^2 + BQ_{\text{inj}} + C \tag{7.11}$$

7.5.6 Polynomial Modeling of Pressure-injection-rate Relationship

To empirically relate injection pressure to rate, a second-order polynomial fit can be used:

$$P_{\text{Fwh}} = AQ_{\text{inj}}^2 + BQ_{\text{inj}} + C \tag{7.12}$$

This formula is a second polynomial equation, relating the following real-time two measured surface parameters:

- Flowing wellhead injection pressure (P_{Fwh})
- Surface injection rate (Q_{inj})

where A, B, and C can be calculated using several surface injection rates with their associated surface wellhead injection pressure. The different injection rates and their associated injection pressure can be either obtained from recent historical data or by changing the choke settings.

7.5.7 Field Example: Real-time Injection Test (Table 7.11)

The wellhead injection pressure is plotted against the injection rates (Figure 7.14), after which a second-degree polynomial can be generated using any regression tool.

Based on the regression process:

The final equation is: $P_{\text{Fwh}} = 1\text{E} - 6^{*}(Q_{\text{inj}})^2 + 0.021^{*} Q_{\text{inj}} + 611$.

This equation can be used in real time to estimate pressure for any given injection rate, or vice versa, where:

Table 7.11 A typical water injection rate test.

Day	Injection Wellhead Pressure (psi)	Injection Rate (MBD)
Day 1	1,554	21.852
Day 2	1,629	22.863
Day 3	1,493	20.710
Day 4	1,467	20.488

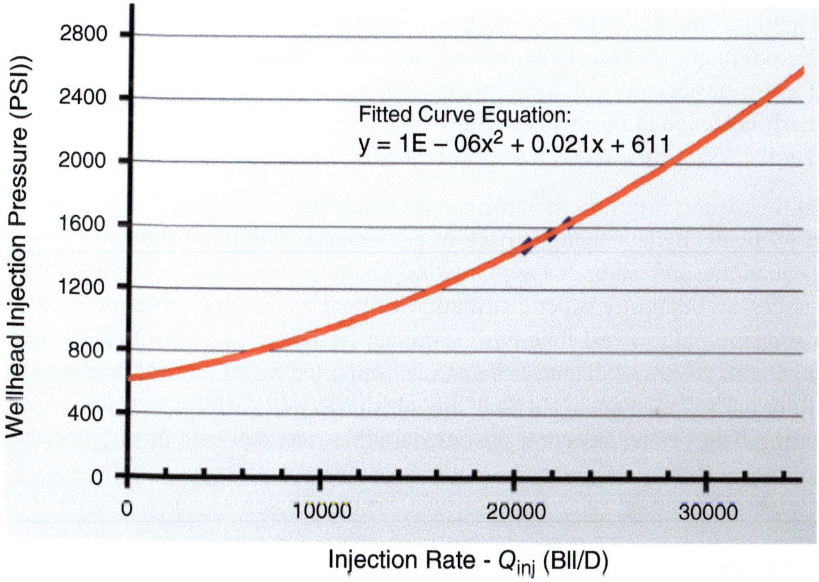

Figure 7.14 Plot of wellhead injection rate versus wellhead injection pressure.

where

- $1E-6=\dfrac{0.000011466 * f * L}{d^5}$

  The fiction factor can be calculated, knowing L and d. The friction factor can also be monitored with time and can help in assessing any scaling or corrosion factor in the injection well.

- $0.021 = 1/\text{II}$, Injectivity Index (II) $= 1/0.021 = 49$ barrel/psi/d

  Changes in II can be tracked over time. For example:

- Year 2002: Measure II = 50 barrel/day/psi
- Year 2007: Measure II = 45 barrel/day/psi

II tracking can be automatically conducted, where stability in II indicates consistent reservoir properties, while declining II suggests possible plugging or reduced permeability near the wellbore due to formation damage.

- $611 = P_{\mathrm{Swh}}$

  Monitoring the real-time changes of wellhead injection pressure can also give an indication of well performance issues. An example of increasing wellhead injection pressure with a stable injection rate can indicate potential negative well performance issues. However, a reduction in the wellhead injection pressure with constant injection rate can indicate either a positive well performance, such as stimulated/fractured formation, or potential downhole leaks.

7.5.8 Applications of Real-time Injection Monitoring

- Optimization of injection pressure and volume in real time.
- Detection of injection anomalies (e.g., formation damage, tubing leaks).
- Field-wide allocation of injection capacity.
- Early detection of injectivity decline.
- Feedback control for automated waterflood systems.

Real-time water injection monitoring and modeling is a powerful capability in smart oil fields. By leveraging surface measurements, simplified generated modeling equations, and calibrated digital twins, engineers can track injectivity, diagnose issues, and optimize water distribution without requiring complex or delayed simulations. Finally, a real-time dashboard can be generated for all field's water injectors with advanced diagnostics analysis supported by AI and ML algorithms to further harvest the benefits of DOF and maximize well performance, leading to increasing field's water injection operational efficiency, reducing downtime, and supporting reservoir pressure maintenance for higher oil recovery.

References

1 Alnuaim S. DOF and reservoir monitoring course notes and presentations. KFUPM; 2020–2025.
2 Alnuaim S. Personal technical insights and field experience in DOF, Smart Wells, SCADA, and reservoir monitoring. Saudi Aramco; 2000–2022.
3 Rassenfoss S. Digital oil field: The future has finally arrived. J. Pet. Technol. 2020;72(6):34–41.
4 Nejad A, Ghorbani S, Davies D. Real-time well integrity monitoring using fiber optic DAS and DTS technologies. SPE Annual Technical Conference and Exhibition; 2019. SPE-196015-MS.
5 McCoy JN, Podio AL. Multiphase Flow Metering Handbook. Elsevier; 2021.
6 Al Khamis MN, Al Khalewi FT, Al Hanabi M, Al Yateem K, Al Qatari A, Al Muailu H. a comprehensive approach of well integrity surveillance, January 19–22, 2014, IPTC-17465-MS.
7 American Petroleum Institute (API). API RP 90-2: Annular Casing Pressure Management for Offshore Wells. API; 2021.
8 McCoy JN, Podio A. Temperature and Flow Diagnostics in Downhole Monitoring. In: Multiphase Flow Metering Handbook. Elsevier; 2021.

9 Al-Masri A, Al-Shalabi EW. Intelligent Well Completions and ICV Optimization: Field Learnings. SPE Kingdom of Saudi Arabia Annual Technical Symposium. SPE-195480-MS; 2019.

10 Chaudhary A, Ghori A, Shoham O. AI-Based Multiphase and Completion Optimization Systems. SPE Annual Technical Conference & Exhibition. SPE-209418-MS; 2023.

11 Beggs DH. Production Optimization Using Nodal Analysis. 2nd ed. OGCI Publications; 2003.

12 Al-Dosari M, Al-Shaibani A. Optimizing multilateral wells using real-time ICVs and digital twins. SPE Paper 194076. 2019.

13 El-Sayed A, Onyekonwu M. Autonomous inflow control using machine learning. J. Petrol. Sci. Eng. 2022;210:110164.

14 Al-Khelaiwi F, Davies D. Intelligent completions—Past, present, and future. SPE Paper 96507. 2005.

15 Brown KE. The Technology of Artificial Lift Methods – Volume 2A: Electric Submersible Pumps. PennWell Books; 2017.

16 McCoy JN, Podio AL. ESP Performance Diagnostics and Multiphase Behavior. Elsevier; 2021.

17 Aslam S, Reddicharla S. AI/ML-Based ESP Failure Prediction Enhances Uptime. J. Petrol. Technol. 2023;75(10):56–61.

18 Kragas T, Hodder M, Weijers L, et al. Autonomous Optimization in ESP Systems Using AI. SPE Artificial Lift Conference & Exhibition. SPE-206322-MS; 2022.

19 Aslam M, Reddicharla V, Mahmoud M. ESP failure prediction using machine learning: ADNOC case study. SPE Paper 212345. 2023.

20 Al-Hashmi A, Al-Kindi S, Al-Busaidi K. Improving ESP run life using real-time surveillance and analytics. SPE Paper 180026. 2016.

21 Garcia M, Smith B. Predictive maintenance of ESP systems using AI: A field demonstration. IEEE Trans Ind Electron. 2021;68(7):5905–5915.

22 Ursini F, Rossi R, Castelnuovo L. The Benefits of Virtual Meter Applications on Production Monitoring and Reservoir Management. SPE 196654, Sep 2019.

23 Ghahremani M, Shoham O, Kouba G. Virtual flow metering using AI/ML: A comprehensive review. SPE Prod. Oper. 2021;36(4):740–58.

24 Perkins TK, Gonzalez M. The Use of Choke Correlations (Gilbert, Ros, Baxendell) for Flow Modeling. SPE Production Engineering. 2018.

25 McCoy JN, Podio A. Empirical and Polynomial Methods in Flow Estimation. In: Multiphase Flow Metering Handbook. Elsevier; 2021.

26 Dake LP. Fundamentals of Reservoir Engineering. Elsevier; 2015.

27 Beggs DH. Nodal Analysis Integration with IPR/VLP Models. In: Production Optimization Using Nodal Analysis. 2003.

28 Al-Mansour M, Dyer S. Evaluation of virtual flow meters vs. MPFMs in brownfields. SPE Paper 181994. 2016.

29 Bikmukhametov T, Jäschke J. First principles and machine learning virtual flow metering: A literature review. J. Petrol. Sci. Eng. 2020;184:106487

30 Temizel C, Duru O, et al. Deep learning applications for reservoir characterization and production forecasting. J. Petrol. Sci. Eng. 2021;205:108841.

31 Al-Ghawas E, Al-Majed A, Al-Kaabi A. DOF-Based Water Injection Surveillance: Field Case Studies. SPE Reservoir Characterisation & Simulation Conference. SPE-197218-MS; 2019.
32 Shahri A, Jaffal W. Real-time injection monitoring and injectivity diagnostics using DOF platforms. SPE Reservoir Eval. Eng. 2022;25(3):601–15.
33 Beggs DH. Polynomial-Based Injection Modeling Methods. In: Production Optimization Using Nodal Analysis. OGCI Publications; 2003.
34 Al-Qarawi A, Al-Jubair A. Real-time monitoring of water injectors using simplified polynomial models. SPE Paper 200232. 2020.

8

Enterprise DOF Workflows—Reservoir Monitoring and Well Testing

Achieving Real-time Reservoir Management Vision

Digital oil field (DOF) technology can be integrated with AI and ML to reach the vision of intelligent real-time reservoir management, helping the industry to harvest every drop of oil and every molecule of gas, leaving absolutely nothing unrecovered. This chapter will focus more on workflows that significantly transformed the way the industry manages the reservoir to ensure healthy reservoirs and maximize recovery.

8.1 Real-time Reservoir Management

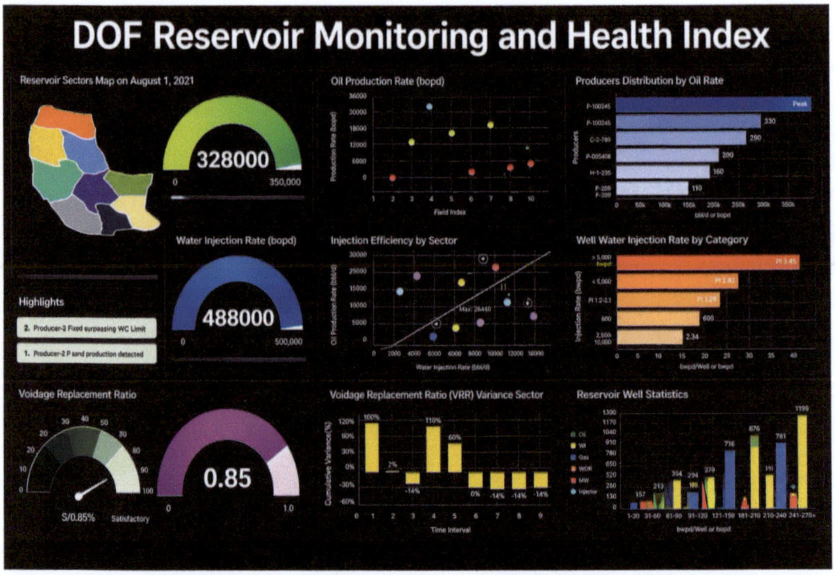

Typical Generic Real-time DOF-generated Reservoir Monitoring and Health Index Charts

The goal of reservoir management is to maximize hydrocarbon recovery while maintaining long-term reservoir health and cost efficiency. Traditionally, this involved periodic well testing, dynamic/static models, and manual analysis. Today, the combination of DOF infrastructure, high-performance computing (HPC), and artificial intelligence (AI)–enabled real-time monitoring, dynamic modeling, and data-driven decisions that significantly enhance reservoir performance. Leading NOC and IOC operators have integrated real-time surveillance with AI technologies to boost recovery, extend reservoir life, and optimize production across complex assets [1–8].

8.1.1 Key Components of Real-time Reservoir Management

Modern DOF reservoir management practice integrates the following core components (Table 8.1).

8.1.2 Real-time Data Sources in DOF Framework

Table 8.2 shows typical reservoir management DOF real-time data that can be further used to improve reservoir analysis.

Major IOCs and NOCs develop a reservoir monitoring and management of DOF platforms/dashboards to address reservoir health, recovery factor, and best reservoir management practices in real time, complemented by advanced analytics and AI and ML algorithms.

Table 8.1 Typical DOF real-time reservoir management components and functionality.

Component	Functionality
Real-time reservoir surveillance	Continuous reliable, QCed, and validated data flow from wells and facilities
AI/ML-based predictive analytics	Forecast reservoir performance, identify anomalies
Reservoir simulation updates	Integrating DOF and HPC, companies have the capabilities to update models in near-real time using history matching
Reservoir management optimization workflows	Recommend well settings, production targets, injection strategies, lift methods, and reservoir health indices
Integrated decision platforms	Enable multidisciplinary collaboration through integrated systems, collaboration centers and shared life dashboards

Table 8.2 Typical reservoir-management DOF real-time data.

Data Stream	Measurement Purpose
Flowing bottom-hole pressure from PDHMS	Used to calculated downhole reservoir static bottom-hole pressure (SBHP) and track reservoir pressure depletion, fluid movement, and ultimate recovery
Multiphase flow rates	Quantify production of oil, water, and gas from individual wells
Injection/production allocation	Monitor recovery processes (e.g., waterflood, CO_2 injection)
DTS fiber-optic sensors	Detect flow profiles, water breakthroughs
Surface network conditions and simulation (digital twin)	Capture optimized choke settings, line pressure, and separator output
Subsurface smart well completion data stream	Zone pressure, temperature, density, ICV settings, and so on are data needed to address subsurface reservoir monitoring, reservoir management, recovery, depletion, sweep efficiency, and zone allocations

8.1.3 AI/ML Applications in Reservoir Management

AI enhances DOF-enabled reservoir management applications by learning from historical and real-time data to guide interventions and ensure healthy reservoirs. The following are good examples (Table 8.3).

8.1.4 Reservoir Health Index Monitoring and Surveillance

Continuous monitoring ensures that the reservoir remains within healthy operating conditions (Table 8.4).

Table 8.3 Typical AI/ML DOF reservoir management-based applications.

Use Case	Description
Reservoir behavior forecasting	Predict pressure decline, fluid contacts, coning behavior
Production optimization models	Adjust lift, choke settings to maximize drawdown efficiency, sweep efficiency and ultimate recovery
Water breakthrough prediction	Classify wells at risk using machine learning
Injection optimization (smart waterflood)	Dynamically allocate injectors based on AI-optimized sweep efficiency
Real-time history matching	Autocalibrate reservoir models as new data streams retrieved from field
Reservoir compartmentalization detection	Use AI to detect unexpected flow barriers or high-permeability streaks

Table 8.4 Typical DOF-based reservoir management health index KPIs.

KPI	Target Behavior
Pressure depletion rate	Controlled drawdown to avoid compaction
Gas–oil contact movement	Slow progression toward producers
Water–oil contact (WOC)	Early warning of breakthrough
Injection efficiency (IE)	>80% in optimized zones
Sweep efficiency	Maximized across injector/producer alignments
Reservoir pressure	Maintain good reservoir pressure above bubble point pressure (BPP) for undersaturated reservoirs and prevent pressure sinks and/or overpressurized zones that would affect the localized and overall recovery factor
Production/injection ratio	Real-time rate compliance and injection/production ratio and drainage

Each application is calculated using real-time data collected from DOF sensors, complemented by required static and historical data and then presented in real-time live reservoir health index dashboards. This is compared to the conventional way of calculating the reservoir health index that used to take weeks and months through lengthy and complex manual methods.

8.1.5 Dynamic Model Updating and Real-time Simulation

AI-enabled workflows now allow near real-time updates to recently history-matched reservoir simulation models by ingesting new production/injection data, recalibrating permeability and porosity zones, and predicting future states.

Digital Twin Reservoir Models

- Live link between field sensors and simulation platforms.
- Enable "what-if" forecasting based on real-time well and injection settings.
- Used for automated scenario planning (e.g., water cut [WC] surge, gas cusping, and water coning).

8.1.6 Optimization Workflows

AI-driven optimization engines assist in the overall reservoir and field:

- Choke management to reduce drawdown stress and well control water coning and gas cusping.
- Well prioritization for stimulation or shut-in.
- EOR agent scheduling (e.g., surfactants, CO_2, and polymers).
- Reservoir balancing via injection rate recommendations.

These are presented through automated dashboards and alerts, allowing reservoir engineers to focus on decision-making rather than manual analysis.

8.1.7 Typical Industry Implementation

Pioneering an automated real-time reservoir management through a comprehensive and integrated DOF platform combines the following:

- Smart wells and intelligent completions using ICVs, DTS, MPFM, and so on.
- High-resolution permanent downhole gauges (PDHMSs).
- Digital twins of gigascale reservoir and surface networks.
- AI-based flow allocation/optimization models.

Expected Results

- Increased recovery factor in selected DOF-enabled fields.
- Enabled 24/7 surveillance and auto-adjustment of water injection.
- Identified low-efficiency injectors using AI cluster models.
- Real-time identification and correction of crossflow zones.
- Optimized water injection strategies, reducing WC in certain fields.
- Reducing development cost to sustain its unprecedented historical low global development cost advantage.

Example: Reservoir A with four multilaterals (zones) was managed using DOF technology to monitor 24 × 7 (Figure 8.1).

Figure 8.1 displays the Reservoir Health Dashboard showing four key metrics across different reservoir zones:

1. **Average Pressure (psi)**: Ensuring reservoir drive is adequate.
2. **WC (%)**: Detecting water breakthrough or coning.
3. **Gas–Oil Ratio (GOR)**: Monitoring changes in fluid behavior.
4. **Injection Efficiency (%)**: Assessing the effectiveness of water or gas injection.

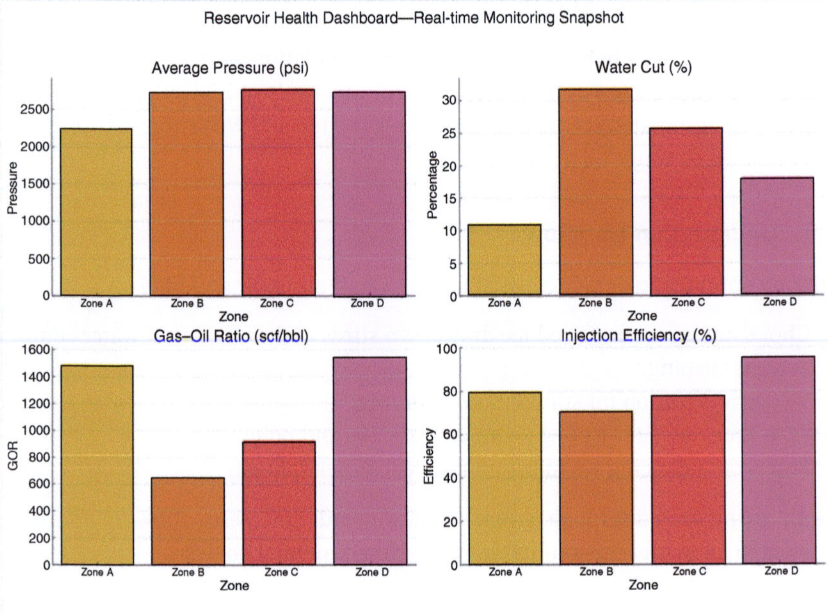

Figure 8.1 DOF real-time reservoir health indices for reservoir A with four zones (generic data).

Each bar graph represents the latest real-time values for a specific zone (A–D), helping reservoir engineers identify early signs of imbalance, optimize injection, and prioritize surveillance or intervention efforts compared to their predefined goals and best practices.

8.1.8 Future Outlook

From NOCs/IOCs intelligent waterflooding, AI-based breakthrough prediction and other smart DOF reservoir workflows, the exciting journey of the oil industry's reservoir management digital transformation is unlocking the full potential of subsurface assets, permanently changing how we manage the earth's most valuable natural reservoirs.

The future of DOF-enabled real-time reservoir management includes:

- Cloud-integrated digital twins accessible to geoscientists worldwide.
- Reinforcement learning to adaptively learn the best reservoir management strategies.
- Satellite/drone-based surface deformation tracking linked to AI subsurface models.
- Edge AI processing for real-time alerts even in remote fields.
- Full real-time dynamic modeling integration with midstream and downstream operations.

Real-time reservoir management is no longer aspirational. The integration of DOF technologies and AI/ML algorithms has enabled unprecedented levels of control, surveillance, optimization, and increasing recovery. By adopting real-time reservoir management strategies, companies are:

- Improving recovery factors.
- Extending reservoir lifespans.
- Enhancing environmental stewardship.
- Ensuring sustainable and more profitable operations.

8.2 Real-time Well Testing in DOF

Well testing has long been a cornerstone of reservoir evaluation and production optimization. Traditionally, this process was time-consuming, equipment-heavy, and delayed by manual data collection and human handling analysis errors. Today, DOF technologies are revolutionizing well testing through real-time data acquisition, wireless telemetry, and cloud-integrated analytics (Figure 8.2) [1–4, 7–11].

This section outlines the evolution, methodology, and advantages of real-time well testing in DOF environments and how it is reshaping subsurface insight, decision-making, and field development.

Table 8.5 shows a traditional versus digital well testing.

8.2.1 Real-time Well Test–Business Need

- Recently developed downhole tools and components controlled remotely via a wireless acoustic telemetry system, and providing real-time access to downhole pressure data has enriched the real-time well testing capabilities.

Figure 8.2 Typical generic DOF real-time well test (AI generated).

Table 8.5 Typical comparison between traditional and digital well test.

Feature	Traditional Testing	DOF-enabled Real-time Testing
Data frequency	Periodic or delayed	Continuous real-time data streams
Data access	On-site and manual	Remote and cloud-based
Processing	Post-test analysis	On-the-fly interpretation
Decision time	Hours to days	Minutes
Personnel required	Large on-site teams, teams and logistics	Minimal with remote support
Well test frequency	Every one to five years	24 × 7 as per need. Can be smartly automated as per the well condition changes, such as WC, FBHP, and GOR

- The evolution of remote connectivity systems and the exploitation of their technical capabilities to their full potential so that off-site well test experts can remotely witness, collaborate, and support well tests.
- The change in the mindset of engineers and operators is essential to ensure that the power of real-time-enabled well testing is fully leveraged. The utilization of real-time well testing should not change the basic concepts.
- Real-time well tests are performed with either multiphase flow meters and/or high-end well test separators for accurate flow rate measurements for better data quality necessary for the well test.
- For conventional well testing data analysis and interpretation purposes, extensive data quality control is carried out after the test is complete, the tools are retrieved, and the data is consolidated.
- Real-time well testing enables continuous quality control and assurance at every step during the test and makes it possible to enact corrective actions when required.
- Bottom-hole pressure is one of the most important parameters acquired during a well test. Hence sensor's reliability and data accuracy are very important to any real-time well test.
- Implementing real-time data quality control (QC), interpretation, and design updates is different than performing these actions once the test is over.
- Time criticality and the lack of complete datasets are significant constraints.
- Effective real-time monitoring requires the ability to handle large quantities of constantly streaming data and workflows that are specifically built for real-time well testing to rapidly convert well test data to meaningful information and make critical decisions.
- Real-time validation of acquired data can mean the difference between a successful test and costly acquisition of non-representative, unusable, or uninformative data.
- Key measurements have to be performed not only accurately but also in a manner representative of the true reservoir response.

Table 8.6 Typical DOF equipment needed for real-time well testing.

Technology	Description
Downhole digital gauges	Capture pressure, temperature, and flow in real time
Wireless telemetry systems	Transmit data from downhole to surface immediately
Surface data acquisition units (DAQ)	Collect, filter, and visualize surface flow data
Cloud integration	Upload test results for remote access and analytics
AI-driven interpretation tools	Automatically classify flow regimes and derive reservoir properties
Digital twins	Simulate test results for comparison and prediction

- Out of the large number of measurements performed during a well test, bottom-hole pressure measurements and production rates can be considered as the most critical.
- The primary validation methodology for pressure measurement consists of comparing measurements from different gauges located at the same point or at nearby points in the production system.

Table 8.6 shows key DOF technologies and equipment needed to conduct real-time well testing.

8.2.2 The Real-time Well Test Workflow

1. **Test Setup**: Permanently Installed downhole gauges and wireless transmitters or deploy intelligent test strings (e.g., memory gauges + fiber-optic sensing).
2. **Data Acquisition**: Pressure transient data, flow rates, and temperature are logged and transmitted continuously to the surface and beyond.
3. **Visualization**: Engineers view pressure buildup/drawdown in real time via dashboards and mobile platforms.
4. **Live Analysis**: Reservoir engineers apply analytical or numerical models using different commercial software, integrated into DOF platforms.
5. **Decision-making**: Data is interpreted on the fly to assess:
   - Productivity index (PI).
   - Reservoir permeability and skin.
   - Boundary effects and interference.
6. **Optimization or Retesting**: If needed, test parameters can be adjusted remotely during the test, avoiding multiple mobilizations.
7. The system can be programmed to conduct a well test after each planned or unplanned well shut-in.

Table 8.7 shows the potential benefits of DOF well testing.

Table 8.7 Potential benefits of DOF-based real-time well testing.

Benefit	Explanation
Faster decisions	Cuts hours or days off standard workflows
Improved reservoir insight	More data points and higher-resolution interpretation
Reduced costs	Minimizes well downtime and eliminates multiple test runs
Remote collaboration	Real-time collaboration between field and office teams
Higher data accuracy	Reduced human error via automated data collection
Advanced analytics	Real-time pattern recognition via AI and ML algorithms

8.2.3 Integration with Other DOF Systems

Real-time well testing is interconnected with broader DOF systems:

- **SCADA Systems**: Real-time control and visibility.
- **AI and ML Platforms**: Detect reservoir anomalies, update simulation models.
- **Digital Twin Systems**: Compare real and simulated behavior to calibrate models.
- **Integrated Asset Management**: Well test outcomes directly inform production planning and field development.

The possible future trends of DOF-based well testing are:

- Wireless downhole tools.
- Edge computing to provide real-time interpretation without cloud delays.
- Autonomous testing systems with AI-driven control of test parameters.
- Nano downhole-sensor arrays for ultra-sensitive measurements for complex reservoirs.

Real-time well testing is one of the most transformative applications of DOF technology. It dramatically improves the speed, safety, and precision of reservoir evaluation while aligning field testing with the broader goals of automation, data integration, and intelligent asset management. As DOF platforms mature, well testing will continue evolving into a real-time, collaborative, and insight-driven discipline.

8.3 Reservoir Pressure Management: Injection/Production Ratio

Effective reservoir management in the era of DOF requires real-time workflows that monitor and optimize injection/production balance, track rate compliance, and maintain dynamic reservoir pressure. This section presents the key intelligent

Figure 8.3 Typical DOF reservoir management workplace and dashboard (AI generated).

workflows—such as injection/production ratio (IPR) monitoring, rate compliance assessment, and automated monthly target setting—that allow engineers to achieve sustainable production while preserving healthy reservoir integrity and pressure (Figure 8.3) [1–4, 8].

The IPR quantifies the volumetric balance between injected and produced fluids in a reservoir or reservoir segment. The IPR process is used to monitor the injection versus production, balancing the reservoir or a segment of the reservoir intake (injection/production). It is an important key performance indicator for operators to monitor the reservoir pressure for that reservoir or segment. Maintaining reservoir pressure is an essential plan in a reservoir depletion strategy, necessary for high recovery factors. The target IPR should be more than the ratio of oil-to-water formation volume factors (B_o/B_w), with some potential assumption of water injection leaking into the aquifer (typically ~10–15%). Your target is to maintain equivalent reservoir voidage to maintain healthy reservoir pressure above the bubble point, especially for undersaturated oil reservoirs.

8.3.1 Key Objectives

- Maintain reservoir pressure above bubble point (undersaturated oil reservoirs).
- Prevent early gas or water breakthrough.

- Match voidage replacement requirements at sector(s) and entire reservoir.
- Account for aquifer leakage (typically 10–15%).

IPR = Q_Injection/Q_Production

- **IPR** must be > $1^*(B_o/B_w)^*1.X$

where

- B_o: oil formation volume factor (RBL/STBL)
- B_w: water formation volume factor (RBL/STBL ~1.0)
- x: aquifer leakage factor (e.g., 0.15)

Example:

If $B_o = 1.3$, $B_w = 1.0$, and $x = 0.15$, then:

$$\text{IPR} = (1.3/1.0)^*1.15 = 1.49{\sim}1.5$$

For a production rate Q_o = 5,000 STBD, injection rate requirement (Q_{inj}) = 5,000 * 1.5 = 7,500 BWD.

8.3.2 IPR Monitoring Workflow (DOF-automated Real-time Workflow)

Steps

1. Divide the field into segments or voidage groups.
2. Source data:
   - Production: MPFM, VFM, PQ curves, or recent test data.
   - Injection: Solar meters, well models (digital twins), or validated test rates.
3. Convert surface volumes to reservoir volumes using formation volume factors.
4. Compute IPR continuously and report monthly.
5. Align production/injection strategy with real-time and simulation-derived pressure targets.
6. Ensure compliance with the target IPR on each reservoir segment as well as the entire reservoir.

This workflow ensures ongoing reservoir balance, supports reservoir simulation models, and informs field development decisions.

8.4 Rate Compliance Monitoring

8.4.1 Typical Chart Showing Rate Compliance of Oil Field

Ensures that individual wells are operating within their assigned targets (e.g., ±5–10%) to ensure full compliance with healthy reservoir voidage, pressure, recovery, and production performance. Figure 8.4 shows a typical DOF generating a real-time chart with different well rate compliance categories (KPIs). Such life charts are monitored in real time by reservoir engineers [1–4, 8].

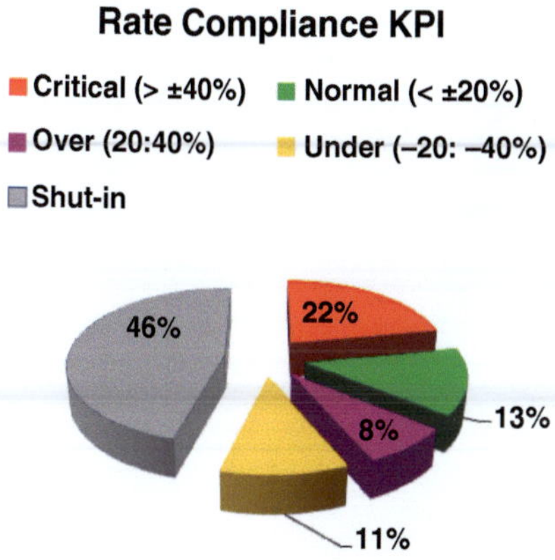

Figure 8.4 Typical real-time DOF rate compliance chart.

8.4.2 Workflow Steps (Using DOF-automated Workflow)

1. Measure real-time rates using MPFM, digital well models, or the latest validated test data.
2. Compare the target rates set by reservoir engineers to the rates measured/calculated in step 1.
3. Calculate the deviation in percentage.
4. Categorize compliance (recommended example—can vary from one company to another):
   - Normal: ±20%
   - Overproducing: +20 to +40%
   - Underperforming: −20 to −40%
   - Critical: ±40%+
   - Shut-in: 0%
5. Visualize using color maps to indicate performance status across the field.
6. Develop a smart alert system triggering real-time tracked alerts to critical non-compliant wells.
7. Link to the overall reservoir health index dashboard with color-rated heat map, complemented by smart alerts and reporting features. This workflow drives operational alignment with strategic production and injection plans.

8.5 Dynamic Reservoir Pressure Monitoring

8.5.1 Typical Generic Isobaric Reservoir Pressure Map

Maintaining healthy reservoir pressure and reservoir pressure above the bubble point (for undersaturated reservoirs) are essential to prevent reservoir damage and to maximize recovery. This can be achieved by generating a real-time reservoir pressure called a reservoir isobaric map or static bottom-hole pressure (SBHB),

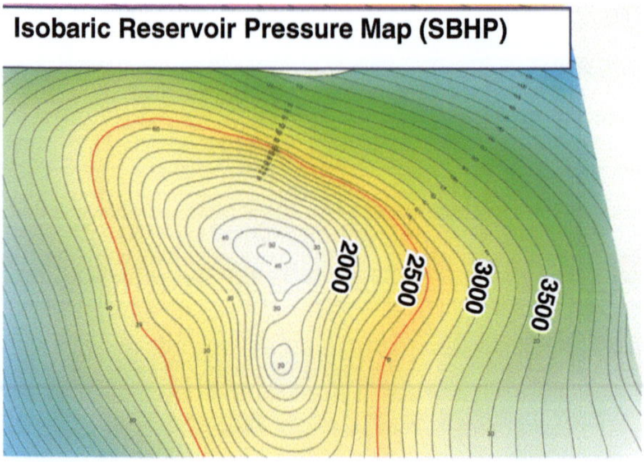

Figure 8.5 A typical real-time live reservoir static bottom-hole pressure (SBHB).

showing the overall reservoir pressure performance, indicating potential problems such as pressure sink (drop) or pressure buildup (Figure 8.5) [1–6, 8, 11].

8.5.2 Workflow Components

- Measure downhole pressure (PDHMS) in producers and injectors.
- Convert wellhead data to sand-face pressure using calibrated well models.
- Use PI/II (productivity/injectivity index) relationships and MPFM measured rate to calculate real-time static bottom-hole reservoir pressure (SBHP).
- Generate real-time pressure maps for continuous surveillance.
- Develop smart analytics that identify pressure sinks and pressure high zones and suggest production/injection optimization to address any anomaly, compared to the set targets.

These colored heat maps offer powerful tools for identifying pressure anomalies and guiding injection management.

8.6 Monthly Production and Injection Strategy

This intelligent workflow automates the generation of well-level targets based on real-time data and calibrated digital well/network models (Figure 8.6) [1–5, 7, 8].

- The workflow is to automate the process of generating reservoir production injection requirements and well targets by integrating real-time data with the latest advances in surface/subsurface facilities modeling (digital well, gas–oil separator plant [GOSP], and field models).
- The deliverable for this workflow is an automated process that can be used to replace a lengthy manual process of generating reservoir production/injection well targets, improving the accuracy, saving time, and accelerating the response to any required change in the field due to oil demand variation.

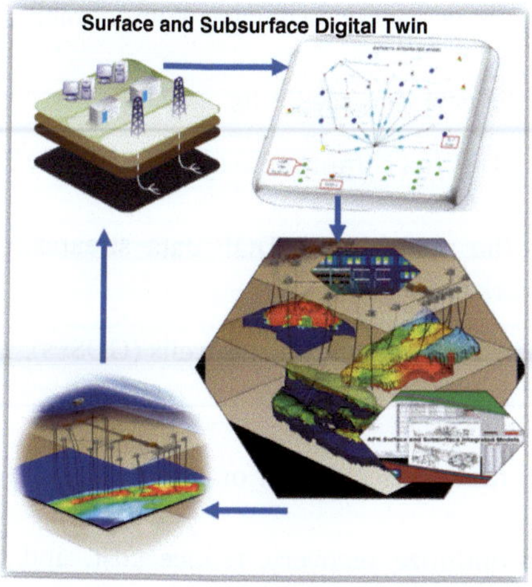

Figure 8.6 Typical surface and subsurface monthly production/injection field model.

- The workflow should be designed to enable engineers to set certain constraints related to reservoir and production facilities to be honored while generating the recommended well(s) production/injection target.
- Constraints can be at the field level, plant, trunkline, group of wells, or even pipeline level.
- Constraints include reservoir drawdown based on recommended IPR, WC, FBHP, pipeline capacity, and GOSP water and gas compression handling capacity.
- The workflow utilizes a calibrated integrated surface and subsurface network model to calculate the recommended production/injection target for each well(s), honoring all provided constraints.
- The workflow can be further used in conjunction with reservoir simulation to calculate the optimum field production injection strategy that would ensure maximum recovery and a longer production plateau.

8.6.1 Workflow Objectives

- Integrate real-time surface/subsurface data from DOF systems and Oracle databases.
- Calibrate digital twins for each well.
- Update GOSP and WIP network models to reflect changing well dynamics.
- Replace manual planning processes with automated optimization tools.

8.6.2 Workflow Constraints and Optimization

The monthly target setting workflow considers multiple field constraints, including:

- Reservoir drawdown (IPR-derived), minimum bottom-hole flowing pressure (FBHP), and max WC, and/or GOR.

- Pipeline, trunkline, and plant capacity (e.g., GOSP water/gas compression handling capacity).
- Group of wells or segment of field-specific performance limits.

These constraints ensure that optimization does not compromise system integrity or exceed facility limitations.

Once all constraints are accounted for, the system uses DOF data streams, integrated with the surface–subsurface network model to:

- Recommend optimal monthly targets for each well, group of wells (GOSPs), and reservoir.
- Allocate injection volumes across zones.
- Adjust production rates as per any operation interruption or oil production demand change during the month.
- Develop short-term strategic plans to optimize recovery, reduce cost, and improve HSE/sustainability performance.

This dynamic model is adaptable, scalable, and can be tied to long-term reservoir simulation for strategic forecasting.

References

1 Alnuaim S. DOF and Reservoir Monitoring Course Notes and Presentations. KFUPM; 2020–2025.

2 Alnuaim S. Personal Technical Insights and Field Experience (35+ years) in DOF, Smart Wells, SCADA, and reservoir monitoring. Saudi Aramco; 1987–2022.

3 Barkved OI, Heavey P, Solbjørg ML. Smart fields and real-time reservoir management. J. Pet. Technol. 2014;66(06):28–35.

4 Saputelli L, Nikolaou M. Smart oil fields: Concepts, benefits, and challenges. SPE Prod. Oper. 2016;31(02):101–114.

5 Oliver DS, Reynolds AC, Liu N. Inverse Theory for Petroleum Reservoir Characterization and History Matching. Cambridge: Cambridge University Press; 2008.

6 Ertekin T, Abou-Kassem JH, King GR. Basic Applied Reservoir Simulation. Richardson (TX): Society of Petroleum Engineers; 2001.

7 Mohaghegh SD. Data-driven Reservoir Modeling. Tulsa (OK): PennWell; 2011.

8 Schlumberger. Digital Oilfield: From Vision to Reality. Houston (TX): Schlumberger; 2017.

9 Horne RN. Modern Well Test Analysis: A Computer-Aided Approach. 2nd ed. Palo Alto (CA): Stanford University; 1995.

10 Bourdet D. Well Test Analysis: The Use of Advanced Interpretation Models. Amsterdam: Elsevier; 2002.

11 Al-Dhafeeri AM, Nasr-El-Din HA. Real-time pressure transient analysis using permanent downhole gauges. SPE Prod. Oper. 2007;22(02):153–161.

9

Enterprise Strategic DOF Workflows

9.1 Surface and Subsurface Facilities Debottlenecking

In the evolving landscape of oil and gas operations, international oil companies (IOCs) are increasingly adopting digital twin (DT) technologies as a central enabler of digital oil field (DOF) strategies. These virtual representations of physical assets—ranging from reservoirs to surface production facilities—allow engineers to simulate, analyze, optimize operations, and debottleneck surface/subsurface facilities in real-time.

The integration of surface and subsurface DTs creates a unified model that mirrors the behavior of the entire production system. This integration enables debottlenecking studies, helping operators identify constraints, optimize flow, and enhance hydrocarbon recovery while maintaining system integrity and minimizing operational costs.

9.1.1 Concept of Integrated DT in DOF

A DT is a dynamic, data-driven replica of a physical system. Within a DOF environment, the subsurface DT models the reservoir, wells, and completion behavior, while the surface twin simulates the processing network, including pipelines, separators, compressors, and storage systems. Integration between these two twins forms a "system-of-systems" model that continuously updates via real-time data feeds from sensors, SCADA, and historians [1–7]. The result is a powerful virtual environment capable of:

- Predicting production under various operating scenarios.
- Identifying bottlenecks across surface and subsurface boundaries.
- Supporting optimization and what-if analyses.
- Developing field production/injection strategy with specific WC, gas–oil ratio, and other operational constraints.

9.1.2 Workflow of a DT Debottlenecking Study

Step 1: Data Integration and Model Building

- **Subsurface Inputs**: Reservoir models/DT, well test data, reservoir simulation outputs, completion design, artificial lift parameters, and PVT data.
- **Surface Inputs**: Production network schematics/DTs, flowline characteristics, separator and compressor capacities, and choke settings.
- **Integration Layer**: Data flows from the DOF's real-time historian into an integrator surface/subsurface DT modeling platform (this could be company-specific middle layer, provided either internally or by one of the SOCs).

Step 2: Model Calibration and Validation

The DT is continuously calibrated against historical and real-time measurements to ensure predictive accuracy. Model validation includes wellhead pressure matching, network pressure drop verification, and fluid composition calibration.

Step 3: Constraint and Bottleneck Identification

Using the integrated model, engineers perform sensitivity analyses to identify:

- Reservoir limitations (e.g., pressure depletion zones or low permeability areas).
- Wellbore restrictions (e.g., ICV settings, sand production, scaling).
- Surface bottlenecks (e.g., separator pressure constraints, compressor surge zones, or pipeline friction losses).

Step 4: Optimization and Scenario Analysis

Debottlenecking scenarios are simulated digitally to evaluate improvement opportunities, such as:

- Adjusting choke settings or ICV positions.
- Reallocating gas lift volumes.
- Modifying separator pressure settings.
- Rerouting flowlines or installing parallel lines.
- Introducing new compression stages or artificial lift upgrades.

Step 5: Implementation and Continuous Monitoring
Once an optimized scenario is selected, real-time data continues feeding the DT to monitor performance post-implementation. Predictive analytics and artificial intelligence (AI) modules detect deviations, prompting proactive maintenance or control actions.

9.1.3 Integrated DOF and DT Benefits

The oil industry literature is full of good examples, showing the advantages of using DOF and DT case studies with significant benefits, including:

- Increasing total oil production without additional CAPEX expenditure.
- Reducing system pressure drop to enhance the overall field's energy efficiency.
- Reducing downtime through predictive monitoring of surface/subsurface equipment.
- Identifying hidden opportunities in constrained wells.
- OPEX reduction, including energy optimization.
- Sustainability gains through lowering flaring rates and reducing power consumption.
- Decision agility through real-time collaboration between surface and subsurface teams.

With it, the DOF becomes a responsive, learning, and optimizing system. It provides advanced capability of identifying bottlenecks before they impact production, optimizing asset performance continuously, and maintaining operational integrity throughout the asset's lifecycle.

The next generation of DTs will feature:

- AI-driven adaptive twins that autocorrect models using machine learning (ML).
- Cloud based integration with corporate planning systems.
- Autonomous control loops where optimization actions (like choke adjustments) are executed automatically within safe limits.
- Sustainability metrics integration, linking debottlenecking improvements to carbon intensity reductions.

Ultimately, the fully autonomous DOF will combine the physical and digital worlds seamlessly, enabling predictive, optimized, and sustainable energy production.

9.2 DOFs and Reservoir Well Placement (Geo-steering)

In the upstream value chain, drilling and well placement are among the most capital-intensive and technically challenging data collection and integration activities. Every decision—from selecting a drilling target to determining the precise trajectory of a wellbore—has a direct impact on development cost, production

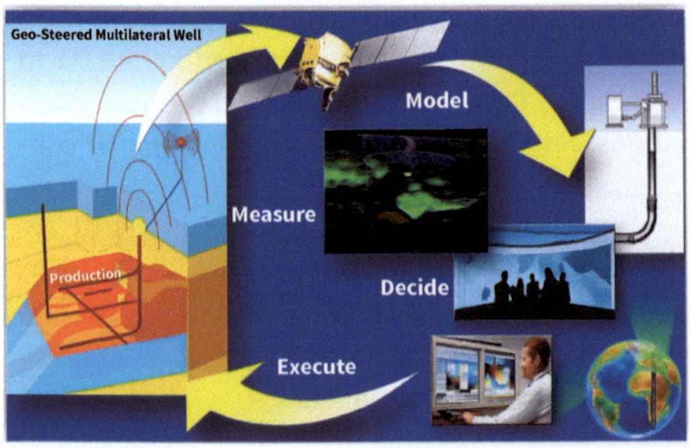

Figure 9.1 Typical well placement DOF workflow.

performance, and ultimate recovery. In this context, the DOF plays a transformative role by bringing together real-time data, advanced analytics, geoscience models, and ML algorithms to enable smarter, faster, and more precise well placement decisions (Figure 9.1) [1, 2, 8–11].

This section explores how DOF capabilities improve the entire lifecycle of well planning and drilling—from prospect evaluation and trajectory optimization to real-time drilling execution and post-drill analysis. Furthermore, it will demonstrate how intelligent field technologies help lower development costs, increase efficiency, and maximize recovery through better well placement and reservoir contact.

9.2.1 Strategic Importance of Smart Drilling and Well Placement

Drilling and well placement are central to field development economics. Errors in well trajectory, poor reservoir targeting, or inefficient drilling execution can lead to:

- Suboptimal production rates.
- Premature water or gas breakthrough.
- Wellbore stability issues and nonproductive time (NPT).
- Increased rig time and drilling costs.
- Missed reserves and stranded hydrocarbons.

A DOF mitigates these risks by integrating data-driven tools into the planning, execution, and optimization stages of drilling and reservoir development.

9.2.2 DOF Capabilities in the Well Planning Phase

The DOF enhances well planning through multidisciplinary collaboration, data integration, and simulation tools. Key capabilities include:

- 3D seismic interpretation, geological modeling, and reservoir modeling integrated with real-time data to accurately identify sweet spots.
- AI-assisted geological uncertainty analysis to reduce risk in well target selection.

- Automated well trajectory planning that maximizes reservoir contact while avoiding faults and hazards.
- Probabilistic drilling cost and risk models for scenario-based decision-making.
- Cross-discipline collaboration platforms allowing geoscientists, reservoir engineers, and drilling teams to align in real time, hence achieving their overall production; health, safety, and environmental (HSE); and financial objectives.

These tools improve well placement precision, reduce planning time, reduce the number of exploratory/development wells, and increase confidence in development strategies.

9.2.3 Real-time Drilling Optimization in the Smart Field

Smart field infrastructure provides real-time visibility into every aspect of the drilling process, enabling faster responses to downhole conditions and reducing unplanned events. Capabilities include:

- Drilling automation systems using AI to adjust weight-on-bit, rotation per minute (RPM), and mud parameters for optimal rate of penetration (ROP) and wellbore cleaning and faster drilling rate (feet/minutes).
- Real-time geo-steering guided by logging while drilling (LWD) data and predictive models.
- Stuck pipe and kick prediction models using ML to detect early signs of trouble.
- Automated rig activity recognition to identify invisible lost time and improve operational efficiency.
- Edge computing to enable rapid decision-making even in remote or offshore environments.

These enhancements reduce NPT, improve safety, and minimize drilling costs.

9.2.4 Intelligent Reservoir Navigation and Geo-steering

Wellbore placement within the optimal zone is critical for maximizing recovery and avoiding early water or gas breakthrough. DOF technologies support this with:

- Dynamic reservoir models updated in real time with LWD and pressure data.
- AI-enhanced geo-steering tools that interpret formation data to adjust well paths during drilling.
- Borehole imaging and ML to classify lithology and fluid contacts.
- Trajectory optimization algorithms to increase horizontal section length within the most productive zones.

The result is improved reservoir exposure, better inflow performance, and enhanced sweep efficiency.

9.2.5 Cost Reduction Through Digital Workflows

The adoption of DOF tools directly contributes to development cost reduction, including:

- Shorter well planning cycles using integrated DTs and simulation.
- Fewer sidetracks and dry holes due to accurate targeting.
- Less rig time through real-time drilling optimization.
- Lower HSE costs by minimizing manual intervention and risk.
- Improved resource allocation through predictive modeling and scenario analysis.

Cumulatively, these efficiencies can significantly cut development costs, depending on the complexity of the asset.

9.2.6 Enhancing Recovery Factors

Smart and well-placed wells lead to higher recovery factors by:

- Targeting high-quality reservoir zones with higher pressure and permeability.
- Optimizing horizontal placement to delay water and gas breakthrough.
- Enabling adaptive completion and injection strategies.
- Increasing the effectiveness of enhanced oil recovery (EOR) schemes through better subsurface understanding.

Fields with DOF-enabled drilling and placement strategies consistently report recovery factor improvements, contributing to significant uplift in reserves and asset value.

DOF platforms and geo-steering are proven tools in the drilling optimization in most major NOCs and IOCs. Referenced case studies show meaningful improvements in:

- Drilling time reduction depending on section and complexity.
- Cost savings.
- Improved wellbore placement (avoiding poor zones and maintaining net pay).
- Operational risk reduction.

9.2.7 Future Outlook: Autonomous Drilling and AI-driven Well Design

Looking ahead, DOF technologies will evolve to support:

- Fully autonomous drilling systems capable of adjusting all parameters with minimal human oversight.
- AI-generated well designs based on continuous learning from historical and real-time data.
- Closed-loop optimization integrating drilling, completion, and production in a single digital framework.
- Next-generation of geo-steering technologies powered by deep learning and 4D reservoir visualization.
- Collaborative design platforms where human and machine cocreate optimized development plans.

These advancements promise to reshape the economics of exploration and development, making even marginal fields viable. DOF technologies are revolutionizing drilling and reservoir development by making every well smarter, safer, faster, and more productive. From enhanced planning to autonomous execution, these tools enable precise well placement, reduce drilling costs, and unlock higher recovery from complex reservoirs. By integrating DOF capabilities into drilling workflows, operators can shift from reactive problem-solving to proactive/predictive development—a critical shift in an era where efficiency, sustainability, and performance come together. The intelligent well is not just a better well—it is the key to a better field, a better business, and a more sustainable energy future.

9.3 DOF Operational Efficiency, HSE, and Sustainability Performance

The oil and gas industry stands at a crossroads: it must continue to deliver reliable, affordable energy while also addressing increasing global pressure to reduce environmental impacts. In this dual-mandate world, smart and intelligent fields are uniquely positioned to deliver operational excellence and support sustainability objectives simultaneously (Figure 9.2) [1, 2, 12–15].

Smart field technologies not only optimize production and reduce costs, but they also minimize energy consumption, emissions, and environmental risks. In this section, we will examine how digital transformation supports both efficiency, HSE, and sustainability responsibility, highlighting specific capabilities, metrics, and case applications that demonstrate how intelligent fields enable sustainable, high-performance operations.

Figure 9.2 Typical balance between operational efficiency and environmental sustainability in oil fields (AI generated).

9.3.1 Redefining Operational Efficiency

Traditionally, operational efficiency in oil and gas focused on maximizing output while minimizing cost and downtime. In smart fields, efficiency is redefined through a more holistic lens, incorporating:

- **Resource Productivity**: Producing more hydrocarbons per unit of energy intensity, water consumption/disposal, emissions, carbon footprint, and labor.
- **Asset Utilization**: Maximizing uptime and throughput of wells, compressors, and facilities.
- **Process Optimization**: Continuous real-time adjustment of production and processing conditions to operate at peak efficiency using real-time data and AI models developed from historical data.
- **Digital Workflow Integration**: Reducing redundancy and manual intervention through automation and the use of drones and robots.
- Integrated corporate upstream, midstream, and downstream digital operations with dynamic real-time overall system optimization linked to global energy supply/demand.

By leveraging real-time data, predictive analytics, and closed-loop control systems, smart fields enable faster, data-informed decisions and minimize human error.

9.3.2 Environmental Sustainability in the DOF

DOF technologies contribute directly to environmental sustainability by:

- Reducing greenhouse gas emissions through optimized energy use and flaring minimization.
- Lowering freshwater use in drilling, stimulation, and waterflood operations.
- Detecting and containing leaks more rapidly to protect ecosystems.
- Optimizing logistics to cut vehicle and equipment emissions.
- Enabling real-time emissions tracking and reporting for compliance and ESG transparency.

Smart fields support not just operational performance but also corporate climate targets and stakeholder expectations.

9.3.3 Emissions Monitoring and Reduction

Methane and CO_2 emissions are major contributors to the oil and gas sector's climate impacts. Smart field technologies enable real-time 24 × 7 monitoring, reporting, and proactive emissions management through:

- Continuous emissions monitoring systems (CEMS) on flares, vents, and compressors.

- Remote sensing via drones, satellites, and fixed sensor arrays.
- Predictive analytics to prevent events that lead to excessive flaring or venting.
- Smart combustion control to optimize burner efficiency and reduce carbon intensity.
- Real-time emissions dashboards integrated with production data.

Companies can now track carbon intensity per barrel in real time and make operational decisions that minimize environmental harm.

9.3.4 Energy Efficiency and Fuel Optimization

Energy is a significant cost and emissions driver in oil and gas production. Smart fields improve energy efficiency by:

- Monitoring and adjusting compressor load to reduce overconsumption.
- Using AI to balance power demand across pumps, drives, and generators.
- Identifying underperforming equipment that wastes fuel or runs below optimal conditions.
- Deploying variable frequency drives (VFDs) controlled by real-time analytics.
- Reduce the number of exploratory wells and the number of required drilled wells to meet a certain production target.
- Implementing smart energy management systems that align operations with off-peak grid pricing or renewable integration.

By improving the energy efficiency of field operations, companies reduce both costs and carbon footprints.

9.3.5 Water Management and Environmental Protection

Water usage, particularly in drilling and enhanced oil recovery, poses sustainability and regulatory concerns. Smart fields address water stewardship through:

- Real-time water tracking systems for consumption, recycling, and disposal.
- Automated leak detection in pipelines and storage systems.
- Predictive corrosion monitoring to prevent environmental damage from pipeline failure.
- Smart injection control to minimize overuse of water in waterflooding or steam injection.
- Early spill detection using sensors and ML analysis of flow anomalies.

These measures reduce environmental impact, support regulatory compliance, and build public trust.

9.3.6 Regulatory Compliance and ESG Reporting

Environmental regulations are tightening worldwide, and investors increasingly expect transparent environmental, social, and governance (ESG) reporting. Smart fields support compliance by:

- Automating data collection and reporting for emissions, discharges, and waste.
- Maintaining digital records for auditing and regulatory inspections.
- Providing near real-time reporting to meet regional or international standards.
- Enabling automated alerts for regulatory threshold exceedance.
- Integrating ESG performance into operational dashboards for cross-functional visibility.

Digital tools not only reduce compliance risk but also enhance an operator's ESG credibility with regulators, investors, and the public.

9.3.7 Circular Operations and Carbon Management

Beyond reduction, the industry is also exploring circularity and carbon management using smart field technologies:

- **Carbon Capture, Utilization, and Storage (CCUS)**: Real-time monitoring of CO_2 injection, storage site integrity, and plume migration through DTs.
- **Produced Water Reuse**: Smart systems manage treatment and reinjection of produced water to reduce freshwater demand.
- **Waste Minimization**: AI-enabled process control reduces chemical overuse and material waste.
- **Lifecycle Analysis Tools**: Assess carbon and environmental footprints of operations end-to-end.

These tools align upstream operations with circular economy principles and long-term climate goals.

9.3.8 Enhancing HSE in Oil and Gas Fields Using DOF Technologies

HSE performance is a critical concern in the oil and gas industry, where operational risks could result in catastrophic consequences. Through DOF technologies, companies now have powerful tools to predict, monitor, and prevent HSE incidents with greater accuracy and speed than ever before.

This subsection explores how DOF systems improve HSE outcomes through real-time awareness, predictive analytics, and automation, with case studies and measurable results from leading energy companies. The data used were all from publicly published documentation.

Table 9.1 shows typical key DOF Technologies contributing to excellence in HSE.

DOF-Automated Driven HSE Feedback Loop: [Sensor Detection] → [Real-Time Alerts] → [DT Simulation] → [Preventive Action] → [Incident Avoidance] → [Better HSE and Sustainability Performance].

Table 9.2 shows typical DOF HSE performance dashboards including DOF platforms' aggregate real-time safety indicators and KPIs.

All KPIs can be filtered by location, time, and shift and integrated into corporate ESG reporting systems.

Table 9.3 shows a typical DOF sustainability and environmental dashboard.

Table 9.1 Typical DOF technologies used in HSE workflows.

DOF Technology	HSE Application
IoT and wearable sensors	Worker tracking, fatigue monitoring, and hazardous gas detection
DTs	Simulation of emergency scenarios, planning safe operations
AI and predictive analytics	Early warning of equipment failure, blowout risk, or leaks
Autonomous drones and robots	Inspection in hazardous zones without exposing personnel
Integrated control rooms	Real-time risk monitoring and incident coordination
Edge computing + cloud dashboards	Rapid response based on field analytics and KPI alerts

Table 9.2 Typical DOF HSE dashboard and KPIs.

Metric	Purpose
Total recordable incident rate (TRIR)	Company-wide safety tracking
Near-miss frequency	Indicates unsafe conditions
Emergency response time	Evaluates readiness and speed
Leak detection response time	Measures containment efficiency
Worker fatigue index	Derived from biometric input

Table 9.3 Typical DOF sustainability and environmental dashboard.

DOF-Driven Initiative	Impact
Methane detection via drone, satellite, and stationary sensors	Meets corporate and/or any oversight authority emissions mandates (e.g., Ministry of Energy)
Real-time spill detection	Enables faster shoreline protection
Smart flaring control systems	Cuts greenhouse gas output
Digital waste management logs	Improves auditability and compliance

Future Trends in HSE Through DOF

- **AI-based Behavior Modeling**: Predict human error before it happens.
- **HSE DTs**: Test emergency plans under various conditions.
- **Voice-activated Smart Personal Protective Equipment (PPE)**: Automated hazard alerts and logs.
- **Drone Swarms for Leaks and Spill Identification and Response**: Fast surface area coverage and surveillance due to:
  - Drone scans pipeline.
  - Sends thermal/methane data to DOF platform.
  - Alert generated instantly on the dashboard.
  - Dispatch automated valve closure if needed.

9.3.9 Example: Sustainability Outcomes from DOF

Saudi Aramco (Source: Saudiaramco.com):

- Saudi Aramco's DOF technologies, such as AI, sensors, DTs, drones, and analytics, played a central role in its excellent sustainability performance. The company has made progress in reducing methane emissions, improving energy efficiency, and strengthening its carbon intensity metrics, even as activity increases. While oil production continues, DOF supports measurable, data-driven improvements across operational and environmental dimensions.
- Saudi Aramco implemented flare monitoring systems (FMSs) and combined heat and power optimization, enabling real-time emissions tracking and scenario-based decarbonization planning. These tools support methane leak detection, flare gas recovery, and energy use forecasting, contributing to reduced greenhouse emissions.
- According to Saudi Aramco's 2024 Sustainability Report, methane emissions declined by 11.4% due to more aggressive flare recovery and digital leak detection efforts; methane intensity dropped from 0.05% to 0.04% of upstream operations.

DOF technologies are reshaping how safety, environmental risk, and sustainability are managed, from reactive conventional to proactive intelligent strategies. With real-time awareness, autonomous monitoring, and predictive intelligence, DOF ensures not only safer operations but also a more environmentally protected and sustainable future for energy. As adoption deepens, corporate HSE and sustainability performance will increasingly depend on how well digital systems are integrated into every layer of oil field operations. As global pressures to decarbonize intensify, operators equipped with intelligent field capabilities will be better positioned to comply, compete, and contribute to a sustainable energy future.

9.4 Data Analytics and ML in DOF

Transforming Oil Field Intelligence from Data to Effective Decisions

The evolution of smart and intelligent fields is deeply intertwined with the rise of data analytics, AI, and ML. While infrastructure—sensors, networks, and computing platforms—forms the physical nervous system of a smart field, data are its lifeblood. By extracting meaning from vast volumes of structured and unstructured data, analytics and ML enable oil and gas operators to move beyond real-time visibility into predictive and prescriptive decision-making [1, 2, 16–19].

In this section, we will explore how data analytics and ML are being applied across the upstream oil and gas value chain. From drilling optimization and reservoir modeling to equipment maintenance and production forecasting, these tools are transforming traditional operations into intelligent, automated systems that continuously learn, adapt, and improve.

9.4.1 The Data Landscape in Oil and Gas

Smart fields generate terabytes of data daily across various domains:

- **Geophysical and Geological Data**: Seismic surveys, well logs, and formation tests.
- **Production Data**: Pressure, flow rates, temperature, water cut, and GOR.
- **Operations and Maintenance**: Equipment condition, failure reports, vibration, and corrosion data.

- **Environmental Data**: Emissions, weather, spill detection, and ambient conditions.
- **Commercial and Financial Data**: Market prices, operational costs, contracts, and logistics.

This diverse data must be ingested, validated, cleansed, and contextualized before it can fuel advanced analytics.

9.4.2 Analytics Maturity Model in Oil and Gas

Smart field analytics typically evolve through a four-stage maturity model:

1. **Descriptive Analytics**: What happened?
   Dashboards, reports, and visualizations based on historical data.
2. **Diagnostic Analytics**: Why did it happen?
   Root-cause analysis using statistical techniques and failure pattern recognition.
3. **Predictive Analytics**: What is likely to happen?
   ML models forecast failures, production, or reservoir behavior.
4. **Prescriptive Analytics**: What should we do about it?
   Optimization algorithms that suggest or automate corrective actions.

The maturity of an asset or operator often depends on the availability of high-quality data and the strength of its data science capabilities.

9.4.3 Examples of ML in Oil and Gas Operations

ML in Drilling Engineering

Drilling operations generate massive real-time data streams from measurement while drilling, LWD, and surface equipment. ML can optimize these complex, high-risk operations:

- **Drilling Parameter Optimization**: Predict the best weight-on-bit, RPM, and mud properties for optimal ROP.
- **Stuck Pipe Prediction**: Classify patterns that indicate impending stuck-pipe incidents.
- **Kick Detection**: Use time-series anomaly detection to identify early signs of formation fluid influx.
- **Automated Rig Activity Classification**: Use computer vision and sensor fusion to track and optimize rig efficiency.

These models reduce drilling NPT, improve safety, and lower costs.

ML in Reservoir Characterization and Modeling

ML enhances traditional reservoir engineering and modeling techniques:

- **Seismic Interpretation**: Use deep learning to classify seismic facies and delineate reservoir boundaries.
- **Petrophysical Modeling**: Apply regression and classification algorithms to estimate porosity, permeability, and saturation from well logs.

- **Reservoir Simulation Acceleration**: Replace complex numerical models with ML-based surrogate models for faster scenario analysis.
- **History Matching Automation**: Calibrate models by matching simulated and historical production data using optimization algorithms.

The result is faster, more accurate insights into reservoir behavior and improved recovery strategies.

ML in Production Optimization and Flow Assurance

Production systems are dynamic and sensitive to operational and environmental variables. ML plays a key role in:

- **Real-time Production Forecasting**: Predict well performance based on historical and current operating conditions.
- **Choke and Lift Optimization**: Use reinforcement learning to find optimal control settings.
- **Slug Flow Prediction**: Detect and mitigate flow instabilities in pipelines.
- **Artificial Lift System Monitoring**: Analyze electrical submersible pumps (ESPs) or gas lift systems to forecast failures or optimize efficiency.

Intelligent production models can respond autonomously to changes in well behavior or market conditions, enabling closed-loop optimization.

ML in Predictive Maintenance and Asset Integrity

One of the most valuable applications of analytics in smart fields is predictive maintenance, which helps prevent equipment failures before they occur:

- **Condition Monitoring**: Use sensors and historical data to monitor critical equipment like compressors, turbines, and valves.
- **Anomaly Detection**: Apply unsupervised ML algorithms to detect deviations from normal operating behavior.
- **Failure Prediction**: Train classification models to predict events like seal failures, corrosion, or fatigue in pipelines and rotating equipment.
- **Remaining Useful Life (RUL) Estimation**: Forecast when equipment is likely to fail, enabling proactive intervention.

These tools extend asset life, reduce downtime, and minimize unplanned production interruptions.

ML in Environmental Monitoring and Emissions Management

Analytics and AI are increasingly central to environmental compliance and sustainability goals:

- **Methane and CO_2 Emissions Detection**: Use AI to process data from satellite imagery, drones, and ground sensors.
- **Flaring Optimization**: Apply real-time analytics to minimize unnecessary flaring events.
- **Spill Detection and Response**: Combine remote sensing, sensor data, and ML to detect and respond to leaks.

- **Energy Efficiency Monitoring**: Analyze fuel usage and equipment efficiency to reduce carbon intensity.

Advanced environmental analytics support ESG reporting and help companies align with climate targets.

9.4.4 Challenges in Data Analytics and ML Implementation

Despite the promise, several challenges must be addressed to fully realize the benefits of analytics in smart fields:

- **Data Quality and Availability**: Incomplete or noisy data can compromise model accuracy.
- **Model Explainability**: Black box ML models can be difficult for engineers to trust or validate.
- **Operational Integration**: Models must fit seamlessly into existing workflows and control systems.
- **Scalability and Generalization**: Models trained on one asset may not work on others without retraining.
- **Skill Gaps**: Engineers may lack data science expertise, and data scientists may lack domain knowledge.

Successful implementation requires multidisciplinary teams, strong data governance, and a culture that values experimentation and learning.

9.4.5 Future Outlook: Toward Autonomous Intelligence

The next phase of smart field analytics is autonomous field operations:

- DTs paired with ML will provide real-time simulations and automated decision-making.
- Edge AI will enable local optimization without latency or cloud dependence.
- Causal AI and hybrid models will combine physics-based and data-driven insights for more robust predictions.
- Human–AI collaboration will enhance decision support, allowing engineers to act faster and with more confidence.

As analytics capabilities mature, smart fields will not only monitor and diagnose but predict, prescribe, and act autonomously, delivering on the promise of truly intelligent oil and gas operations. Analytics and ML are not ancillary tools—they are the decision-making engine of the smart field. From drilling optimization to emissions monitoring, they enable operators to anticipate issues, improve efficiency, and maximize value. But realizing their full potential requires more than algorithms—it demands integrated data, collaborative teams, and a willingness to rethink how decisions are made in the oil field.

References

1 Alnuaim S. DOF and Reservoir Monitoring Course Notes and Presentations. KFUPM; 2020–2025.

2 Alnuaim S. Personal Technical Insights and Field Experience (35+ years) in DOF, Smart Wells, SCADA, and reservoir monitoring. Saudi Aramco; 1987–2022.

3 Grieves M, Vickers J. Digital twin: Mitigating unpredictable, undesirable emergent behavior in complex systems. In: Kahlen FJ, Flumerfelt S, Alves A, editors. Transdisciplinary Perspectives on Complex Systems. Cham: Springer; 2017. pp. 85–113.

4 ISO. ISO 23247-1: 2021 – Automation Systems and Integration—Digital Twin Framework for Manufacturing—Part 1: Overview and General Principles. Geneva: International Organization for Standardization; 2021.

5 Bell G, McDonald J, McKinley J. Integrated production system modeling for field debottlenecking. SPE Prod. Oper. 2013;28(03):238–249. https://doi.org/10.2118/161293-PA

6 Sayda A, Monaghan P, Hart D. Closed-loop optimization using integrated surface and subsurface models. SPE Intelligent Energy Conference & Exhibition. Utrecht, Netherlands; 2014. SPE-167823-MS.

7 Haldorsen JBU, Damsleth E, Gringarten AC. Reservoir management and the digital oil field. SPE J. 2016;21(01):1–14.

8 SPE. The Digital Oil Field: Concepts, Technologies and Challenges. Society of Petroleum Engineers; 2014.

9 Aadnoy BS, Cooper I, Miska S, et al. Advanced Drilling and Well Technology. 2nd ed. Houston: Gulf Professional Publishing; 2019. ISBN: 9780128202219

10 Downton G, Browne M, Hartley J. Real-time geosteering using LWD data and integrated digital workflows. SPE Annual Technical Conference and Exhibition. Houston, USA; 2015. SPE-174974-MS.

11 Schlumberger. The digital oilfield. Oilfield Review 2017;29(1):4–17.

12 Shell Global Solutions. Smart Fields: Delivering Value through Digitalization. Shell; 2018.

13 Saudi Aramco. Sustainability report 2024. Dhahran. Saudi Arabian Oil Company; 2024. Available from: https://www.saudiaramco.com

14 International Energy Agency. Methane tracker 2024. Paris: IEA; 2024. Available from: https://www.iea.org

15 IPIECA. Digital Technologies and Environmental Performance in the Oil and Gas Industry. London: IPIECA; 2022.

16 API. API RP 754: Process Safety Performance Indicators for the Refining and Petrochemical Industries. 2nd ed. Washington (DC): American Petroleum Institute; 2021.

17 Qin SJ, Chiang LH. Advances and challenges in process data analytics and machine learning applications. Annu. Rev. Chem. Biomol. Eng. 2019;10:465–488.
18 Mohaghegh SD. Shale Analytics: Data-driven Analytics in Unconventional Resources. Cham: Springer; 2017.
19 Bishop CM. Pattern Recognition and Machine Learning. New York: Springer; 2006. ISBN: 9780387310732.

10

DOF and Corporate Management

Oil and Gas Corporation Executives' DOF Dashboards (Generated with AI using OpenAI).

The energy sector's transition to digitally enabled operations has transformed how oil and gas companies manage performance at all levels. The digital oil field (DOF) represents an integrated ecosystem where physical assets, automation systems, data analytics, and enterprise platforms converge to deliver actionable intelligence for all corporate levels based on their own needs and official roles.

For corporate executive management, the digital oil field provides a powerful high-level tool to monitor real-time corporate key performance indicators (KPIs) aligned with its strategic goals, including total production; corporate health, safety, and environment (HSE); sustainability; and financial performances. By embedding and integrating multisources QCed real-time DOF data with corporate HSE/sustainability/financial data, executives' roles can shift from traditional reactive to smart/intelligent proactive governance.

10.1 The Executive Perspective on DOF

From a leadership standpoint, the digital oil field is not just a technology project, it is a strategic enabler for:

- Operational excellence
- Corporate transparency
- Risk mitigation
- Sustainability and compliance
- Profitability and shareholder value

Executives must have the strategic insights that DOF is a must-have technology investment to achieve all of the previous strategic corporate objectives. It is not a luxury gadget as it used to be during the DOF's early time. However, to enable executives to achieve the above objectives, they require real-time, aggregated, and context-rich dashboards that offer a single version of the truth across global corporate operations, addressing the corporate short-, mid-, and long-term strategic objectives. Real-time DOF data from upstream, midstream, and downstream have to be fully and digitally integrated with supply chain, finance, and other corporate support organizations including HSE [1–6].

10.2 Strategic KPI Domains Monitored via Digital Oil Fields

The following different corporate KPI categories are recommended [1, 2, 7–15]:

- **Operational KPIs**: Table 10.1

Table 10.1 Recommended corporate operational KPIs.

KPI	Description	Executive Impact
Production uptime	% of time assets are producing	Identifies underperformance, supports asset strategy
Planned versus actual production	Daily/monthly output versus target	Forecast accuracy, project deviation alerts
Overall corporate well/field health index	Composite of multiwell/field integrity data sources highlighting major and critical anomalies	Prioritization of OPEX and CAPEX
Maintenance compliance	% of preventive tasks on time	Asset integrity and unplanned downtime reduction
Corporate fields' water cut and GOR trends	Fluid ratios by well/field	Reservoir health and enhanced oil recovery planning

Sample charts:

- 📊 Live production bar chart by asset/region
- 📉 Trend line of actual versus forecast production (total, crude grade and field)
- 🗺️ Global corporate's fields map with production heat zones from all assets
- ⚠️ Deferred production if any (dead wells due to failed ESP, well integrity, etc.)

- **HSE KPIs**: Table 10.2

Table 10.2 Recommended corporate HSE KPIs.

KPI	Description	Executive Impact
Lost-time injury frequency rate (LTIFR)	Injuries per million man-hours	Safety culture and compliance reporting
Near-miss reporting rate	Frequency of reported near-misses	Risk behavior monitoring and proactive training
Gas leak detection events Industrial fires	Real-time detection logs	Environmental incident prevention
Permit to work (PTW) violations	Unapproved or late permits	Safety management and procedural discipline
Emergency drill compliance	Completion rate and time	Operational readiness scorecard
Fatality	Annual corporate-level fatality record	Safety culture and compliance reporting

Sample charts:

- 📈 Trend line of LTIFR over 12 months
- 🔺 Incident bubble chart (severity vs. frequency)
- 🌍 Environmental compliance map
- ⏱️ Live corporate industrial fire and fatality counter
- ⏱️ Live PTW violations counter
- 🔥 Real-time flaring tracker (mcf/d)

- **Sustainability KPIs**: Table 10.3

Table 10.3 Recommended corporate sustainability KPIs.

KPI	Description	Executive Impact
Greenhouse gas (GHG) emissions (scope 1, 2, 3)	Accurate real-time QCed and approved reporting	Corporate sustainability performance, image, and executive-relevant bounce
Carbon intensity (kg CO_2e/boe)	Accurate real-time QCed and approved reporting	Corporate sustainability performance, image, and executive-relevant bounce

(Continued)

Table 10.3 (Continued)

KPI	Description	Executive Impact
Methane emissions rate (%) Flaring volume (MMscf/month) Flaring intensity (MMscf/boe)	Accurate real-time QCed and approved reporting	Corporate sustainability performance, image, and executive-relevant bounce
Energy consumption per unit of production Energy intensity	Accurate real-time QCed and approved reporting	Corporate sustainability performance, image, and executive-relevant bounce
Water usage and reuse (%) Produced water management (reinjection, treatment, discharge volumes)	Accurate real-time QCed and approved reporting	Corporate sustainability performance, image, and executive-relevant bounce
Spill frequency and volume	Accurate real-time QCed and approved reporting	Corporate sustainability performance, image, and executive-relevant bounce
Waste generated and recycled (tons/year) Carbon circular economy	Accurate real-time QCed and approved reporting	Corporate sustainability performance, image, and executive-relevant bounce

Sample Charts:

- ♻️ Emissions funnel chart
- 💧 Water usage vs. target bar chart
- 📈 CO_2 per BOE trendline for scope 1, 2, and 3
- 🏢 ESG compliance scorecards
- 🗺️ Sustainability KPI heatmap by region

- **Financial KPIs**: Table 10.4

Table 10.4 Recommended corporate financial KPIs.

KPI	Description	Executive Impact
OPEX per BOE $/Barrel or $/MSCFT	Operating cost per barrel of oil equivalent	Field economics, cost optimization
NPV/IRR of assets	Projected asset value over life	Portfolio optimization and investment prioritization
Real-time CAPEX tracking	Budget versus actual project spends	Project governance and financial control
Deferred production value	Revenue loss due to downtime	Operational urgency and prioritization

Sample charts:

- 💰 Budget versus actual bar graphs
- 📉 Rolling cash flow forecast line graph
- 📈 OPEX per BOE trend
- 📈 Development/production cost trend
- 📊 Top projects by IRR scorecard
- 📍 Bubble map of financial risk zones

10.3 Digital Infrastructure Enabling Executive Monitoring

10.3.1 Enterprise-wide Integration Architecture

The typical data flow for executive dashboards includes [1, 2, 15–19]:

- Field sensors and SCADA systems → Edge or central data lakes (% of commissioned, connected-to-desktop, % of data accurate).
- Real-time analytics engines → Enterprise performance dashboards.
- Corporate relational databases (ERP) and systems used to manage the corporate core businesses, supply chain and financial systems → Budget and cost data integration.
- Artificial intelligence/machine learning engines → Predictive KPIs and risk management scenario modeling.

Visualization and Access

- **Executive Cockpits**: Web-based and mobile dashboards customized for different business units.
- **KPI Drilldowns**: From corporate level to asset or even individual well performance.
- **Alerts and Escalations**: Automated thresholds and anomaly detection.
- **What-if Simulations**: Decision support based on live data risk management modeling.

10.4 Best Practices for Executive-level DOF Utilization

Table 10.5 shows some recommended DOF utilization best practices for oil and gas companies' executives [1, 2, 20–26].

Table 10.5 Recommended DOF utilization best practices.

Practice	Description
Align KPIs to strategy	Every KPI shown should link to business objectives (e.g., safety, efficiency, profitability)
Define data ownership	Assign accountability for data quality at the field, country, and corporate levels
Build a KPI governance model	Standardize definitions and thresholds across regions and business units
Use predictive and prescriptive analytics	Move beyond monitoring into decision optimization
Ensure cybersecurity and data privacy	Protect sensitive data and meet compliance standards
Adopt a change management framework	Drive cultural shift toward data-driven leadership

10.5 Benchmark Study of Several International Oil Companies

The following are benchmark results collected from several public companies' websites [1, 2, 27–29]:

How CEOs Use These Benchmarks

- **Target Setting**: Align OPEX, emissions, and safety targets to top decile performers.
- **Peer Comparison**: Compare KPIs with similar-sized peers (bands A/B).
- **Performance Flags**: Alerts for KPIs outside benchmark ranges trigger investigation.
- **Reporting**: Validate that ESG disclosures match operational performance.

Here's a visual benchmark comparison of key executive-level KPIs across the oil and gas industry (Figure 10.1).

KPI comparison highlights the top performers, global max, and industry average of the following key KPIs:

- OPEX per BOE.
- Safety (LTIFR and TRIR), CO_2, and methane intensity.
- Production efficiency.

These real-time live benchmarks enable CEOs and top executives to evaluate performance against peers and set realistic targets.

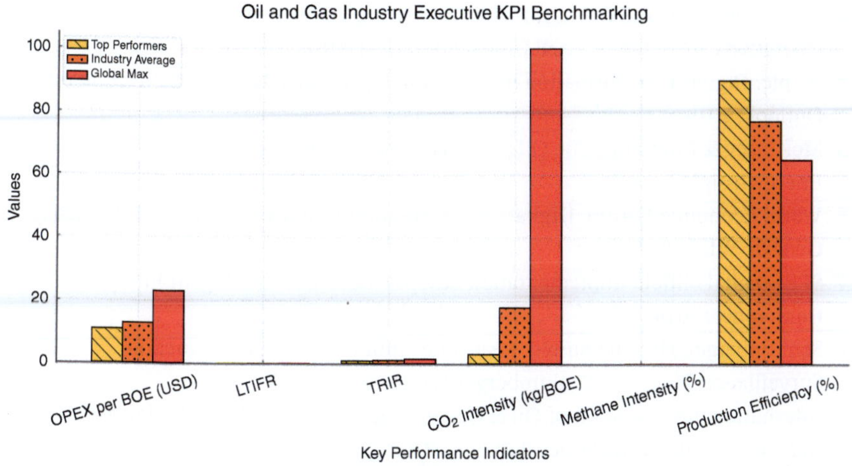

Figure 10.1 Typical oil industry benchmark of key executive KPIs.

10.6 Challenges and Mitigation Strategies

The following are some of the potential challenges and mitigations that corporate executives may face and promptly deal with:

- **Data Silos**: Integrate systems via middleware and data lakes.
- **KPI Overload**: Prioritize strategic KPIs and hide noise.
- **Cultural Resistance**: Embed digital KPIs into performance reviews and incentives.
- **Skill Gaps**: Provide digital literacy training to executives.
- **Legacy Infrastructure**: Use digital twins and edge computing to modernize gradually.

For executive leadership, the digital oil field is more than a technical upgrade. It's a strategic control tower. It transforms how decisions are made, risks are managed, and value is created. When leveraged correctly, DOF enables real-time transparency, faster course correction, and greater alignment between operational and strategic vision [1, 2, 30–32].

References

1 Alnuaim S. DOF and Reservoir Monitoring Course Notes and Presentations. KFUPM; 2020–2025.

2 Alnuaim S. Personal Technical Insights and Field Experience (35+ years) in DOF, Smart Wells, SCADA, and reservoir monitoring. Saudi Aramco; 1987–2022.

3 Shell International BV. Smart Fields: Delivering Value through Integrated Operations. The Hague: Shell Global Solutions; 2016.
4 BP plc. Digital Transformation in Upstream Operations. London: BP Technology Outlook; 2019.
5 McKinsey & Company. The Oil and Gas Organization of the Future. McKinsey Energy Insights; 2018.
6 World Economic Forum. Digital Transformation Initiative: Oil and Gas Industry. Geneva: WEF; 2017.
7 SPE. Digital oilfield implementation best practices. SPE Intelligent Energy Conference; 2014.
8 Schlumberger. Digital Oilfield: Real-time Optimization and Production Surveillance. Houston: Schlumberger White Paper; 2015.
9 International Association of Oil & Gas Producers (IOGP). Safety Performance Indicators – 2023 Data Series. London: IOGP; 2023.
10 International Labour Organization. Lost Time Injury Frequency Rate (LTIFR): Definition and Reporting Guidance. Geneva: ILO; 2018.
11 IPCC. 2019 Refinement to the 2006 IPCC Guidelines for National Greenhouse Gas Inventories. Geneva: IPCC; 2019.
12 IPIECA. Sustainability Reporting Guidance for the Oil and Gas Industry. London: IPIECA; 2021.
13 Global Methane Initiative. Methane Emissions Measurement and Mitigation in Oil and Gas. Washington (DC): GMI; 2020.
14 Damodaran A. Applied Corporate Finance. 4th ed. New York: Wiley; 2015.
15 Society of Petroleum Engineers. Petroleum Economics and Decision Analysis. SPE Monograph; 2013.
16 Gartner Inc. Data and Analytics Architecture for Executive Dashboards. Gartner Research Note; 2020.
17 Microsoft. Azure Architecture for Oil and Gas Digital Transformation. Redmond (WA): Microsoft Industry Solutions; 2021.
18 Amazon Web Services. Oil & Gas Data Lakes and Real-time Analytics Reference Architecture. Seattle (WA): AWS Whitepaper; 2020.
19 NIST. Cybersecurity Framework for Critical Infrastructure. Gaithersburg (MD): National Institute of Standards and Technology; 2018.
20 Kaplan RS, Norton DP. The Balanced Scorecard: Translating Strategy into Action. Boston: Harvard Business School Press; 1996.
21 ISO. ISO 55000: Asset Management – Overview, Principles and Terminology. Geneva: International Organization for Standardization; 2014.
22 PwC. Oil and Gas Digital Strategy and Change Management. PwC Energy Advisory; 2019.
23 Accenture. Digital Dashboards for Oil and Gas Executive Decision-making. Accenture Industry X; 2020.
24 SAP SE. Real-time Executive Dashboards for Oil and Gas Enterprises. Walldorf: SAP Oil & Gas Solutions; 2019.
25 Few S. Information Dashboard Design: The Effective Visual Communication of Data. 2nd ed. Sebastopol (CA): O'Reilly Media; 2013.

26 Tufte ER. The Visual Display of Quantitative Information. Cheshire (CT): Graphics Press; 2001.

27 Wood Mackenzie. Global Upstream Benchmarking Metrics. Edinburgh: Wood Mackenzie; 2022.

28 Rystad Energy. Upstream Cost, Emissions, and Efficiency Benchmarking. Oslo: Rystad Energy; 2023.

29 International Energy Agency. Emissions from Oil and Gas Operations. Paris: IEA; 2022.

30 Deloitte. Managing Data Silos and Digital Risk in Oil and Gas. Deloitte Energy Insights; 2020.

31 MIT Sloan Management Review. Building a Data-driven Organization. Cambridge (MA): MIT; 2019.

32 ISO/IEC. ISO/IEC 27001: Information Security Management Systems. Geneva: ISO; 2013.

11

DOF Future: Toward Fully Autonomous and Sustainable DOF

Beyond Optimization: Charting the Next Frontier of Energy Intelligence (Generated with AI using OpenAI).

The smart and intelligent field is no longer a concept of the future—it is today's operational reality. Companies around the world have begun leveraging real-time data, artificial intelligence, and advanced automation to transform how they explore, produce, and manage hydrocarbon assets. But this evolution is far from complete.

As the energy landscape continues to change—driven by digital acceleration, global climate goals, and economic volatility—the future of oil and gas fields will be defined not just by optimization, but by autonomy, sustainability, resilience, and smart real-time self-management/optimization (fully autonomous intelligent fields). In this closing chapter, we look ahead to the next phase in the evolution of smart fields: one in which human and machine intelligence collaborate to create operations that are self-aware, self-healing, and aligned with the global energy demand, its global environmental goals of net-zero methane and carbon emissions and overall corporations' (IOCs & NOCs) strategic goals.

11.1 From Smart to Autonomous Fields

While today's smart fields offer enhanced visibility and decision support, the next generation will take a significant technological forward leap toward fully autonomous operations—systems that can diagnose and act on their own, with minimal human intervention.

Key characteristics of fully autonomous fields:

- **Self-monitoring**: Sensors and artificial intelligence (AI) models detect changes and anomalies in real time.
- **Self-diagnosing**: Systems identify the causes of operational problems, inefficiencies, risks, and/or failures.
- **Self-optimizing**: Optimization algorithms such as adjusting production parameters to maximize output, reduce cost, and/or minimize emissions.
- **Self-healing**: Maintenance systems predict and automatically schedule repairs or system reconfiguration.
- **Self-learning**: Fully integrated and continuously self-improving systems based on field data, historical patterns, and ever learning/improving field-level AI algorithms.

The autonomous field will function as a closed-loop system, where physical processes and digital intelligence operate in unison with minimum human intervention [1–6].

11.2 The Role of AI, ML, and Complex Analytics

In the coming decade, AI, machine learning (ML), and advanced complex analytics will evolve from decision support tools to a decision-executing system, supported by robots, enabled by:

- Reinforcement learning for continuous process improvement.
- Hybrid AI, ML, and complex analytics models that combine physics-based simulations with ML and digital twins.
- Natural language interfaces allowing engineers to interact with field systems via voice or text.
- Causal inference to understand not just correlations but cause-and-effect relationships in field behavior.
- Federated learning to train AI, ML, and complex analytics models across decentralized assets while preserving data privacy, safety, and cybersecurity.

AI, ML, and complex analytics will become the field's "digital brain," coordinating thousands of sensors, machines, robots, and workflows in real time [1, 2, 7–10].

11.3 The Rise of Industrial Digital Twins at Scale

Digital twins—already used in isolated assets—will evolve into full-field and enterprise-wide live and connected digital twin ecosystems [1, 2, 11–14]. These ultra-detailed virtual models, integrated with more advanced DOF data and more sophisticated AI, ML and complex analytics, will:

- Simulate reservoir dynamics, facility operations, and equipment in real time.
- Run predictive scenarios for production planning, maintenance, and carbon intensity.
- Integrate with midstream, downstream, market, and logistics systems for commercial optimization.
- Serve as training environments for personnel through immersive simulations (VR/AR).

As fidelity and processing power increase, digital twins will become decision engines, not just supporting visual models.

11.4 Real-time ESG and Carbon Circular Economy

As global energy issues such as climate change continue to push the ESG envelope of the oil and gas industry, in such a carbon-constrained world, ESG performance will become as central as production metrics. Hence, the DOF of the future will contribute significantly through:

- Tracking greenhouse gas emissions per barrel/SCFD in real time using embedded sensors and satellite remote sensing feeds.
- Optimizing operations based on carbon price signals or regulatory thresholds, especially when the carbon tax and carbon market become a UN-enforced reality.
- Providing auditable ESG data to investors, regulators, and partners.
- Integrating sustainability into automated workflows and corporate strategies, such as avoiding flare triggers or maximizing energy recovery.
- Enabling circular operations, including closed-loop water systems, CO_2 injection optimization, and overall carbon circular economy concept.

Environmental data will no longer be after-the-fact—it will be live, operational, and actionable [1, 2, 15–18].

11.5 Integration with Renewable Energy and Hybrid Fields

As the world transitions toward a lower-carbon, cleaner energy mix (including cleaner/efficient, low-carbon fossil fuels, renewables, hydrogen, and nuclear), the DOF infrastructure may evolve into a large-scale hybrid energy hub, capable of integrating oil, gas, hydrogen, nuclear, and renewables, including the following:

- On-site solar and wind powering oil and gas fields' compressors and pumps.
- Battery storage managed by AI for load balancing and DOF power optimization.
- Hydrogen and biofuel integration into field logistics and processing.
- Real-time energy source switching based on availability and emissions goals.
- Smart renewables-based microgrids for isolated operations or offshore platforms.
- Geothermal, white hydrogen (produced from natural subsurface hydrogen resources) and in situ subsurface power generation (using fluids) integration in DOF infrastructure.

The result will be more resilient, agile, cost-effective, and sustainable digital energy fields' (DEF) ecosystems that will meet the ever-increasing global energy demand, and the global low-carbon energy mandate, toward unlocking the global energy sustainability dreams of the net-zero CO_2 and methane emissions' stretched targets by 2050/2060 [1, 2, 19–21].

11.6 Quantum Computing and Next-gen Simulations

As datasets grow and models become more complex, quantum computing will play a key role in solving previously intractable problems:

- Reservoir simulations with billions of dynamic variables, surface, subsurface, logistics, and global energy markets (supply and demand).
- Molecular modeling of fluid interactions and enhanced secondary/tertiary oil recovery agents at the nanoscale.
- Optimization problems involving drilling path selection (smart drilling), facility layout, and/or logistics planning.
- Cryptography and data security for ultra-secure communications in critical infrastructure.

Though still in early stages, quantum computing promises to unlock a new level of precision and scale in oil and gas optimization, helping the industry to unlock hundreds of opportunities to increase recovery, reduce development cost, and significantly improve ESG performance [1, 2, 22–24].

11.7 Robotics and Drones in DOF Automation

As the oil and gas industry evolves into a digitally driven ecosystem, robotics is emerging as a transformative force. In digital oil fields (DOF), robotics is not only enhancing operational efficiency and safety but also enabling tasks that were previously considered too hazardous, complex, or costly. From autonomous inspection drones to robotic drilling assistants, robotics is helping reshape the value chain from exploration to decommissioning [1, 2, 25–27].

Although today's DOF started integrating drones and robots into its infrastructures, utilizing its significant benefits, smart fields of the future will be highly populated not just by smart sensors but by autonomous intelligent robots and drones, capable of physically smart and effective fields/facility/well intervention, including but not limited to the following:

- Smart inspection drones using computer vision and thermal cameras for remote asset surveillance.
- Smart crawling robots for pipeline monitoring, corrosion checks, and confined space inspection.
- Smart automated drilling rigs' ecosystems for hands-free drilling and completions operations.
- Smart field robots capable of opening valves, conducting repairs, or cleaning equipment.
- Smart swarm robotics for coordinated, distributed, complex, risky field tasks.

Integration of smart robots and drones, combined with AI and edge computing, will extend DOF's lifespan and its operational efficiency while reducing safety/security risks.

11.7.1 Potential DOF Future of Robotics and Drones

Robots and unmanned aerial vehicles (UAVs), commonly known as drones, are rapidly becoming a critical tool in DOFs. Their ability to provide high-resolution data, real-time monitoring, and access to hard-to-reach areas makes them essential for modern oil and gas operations. From infrastructure inspection to emissions tracking, drones enhance safety, reduce operational costs, and support predictive analytics across the field lifecycle [1, 2, 28–33].

11.7.2 Key Functions of Robots and Drones in DOF

- Inspection and monitoring (e.g., flare stack inspections and pipeline patrol).
- Surveying and mapping (e.g., 3D modeling of construction sites).
- Environmental monitoring (e.g., leaks, methane, and CO_2 emissions detection).
- Emergency response (e.g., spill assessment and heat detection in fires).

11.7.3 Integration with DOF Technologies

Robotics and drone systems are deeply integrated with:

- **IoT**: Robots communicate with IoT sensors for real-time status updates and adaptive behaviors.
- **AI and ML**: Algorithms guide decision-making, such as identifying corrosion levels or predicting failures.
- **Digital Twins**: Drones and robots update virtual replicas of physical assets, enabling simulation-based decision-making.
- **Cloud Platforms**: Storage and processing of robots and drone data integrated with enterprise decision systems.

Benefits

- **Safety**: Reduces human exposure to hazardous environments (DOF pipeline inspection, flare stack surveys, etc.).
- **Efficiency**: Robots work continuously without fatigue, enabling 24/7 operations.
- **Accuracy**: Precision tools and AI guidance lead to better outcomes than manual processes.
- **Cost-Effectiveness**: Lower long-term operational costs despite high upfront investment.

11.7.4 Challenges and Ethical Considerations

Balancing technological progress with global energy sustainability/responsibility will be a defining challenge of the next era. The future is promising—but not without risks and responsibilities:

- **Cybersecurity**: More connected systems mean more attack surfaces.
- **Data Ethics**: Who owns, controls, and benefits from operational data?
- **Workforce Impacts**: How do we reskill and retain value in a changing labor landscape?
- **Technology Governance**: How do we prevent black box AI from making opaque or unsafe decisions?
- **Sustainability Trade-offs**: How do we ensure that digital solutions themselves are energy efficient?

Robotics in DOFs is not a futuristic concept—it's a growing reality driving safety, efficiency, and sustainability in the energy sector. As the industry continues to digitally transform, the integration of robotics will play an indispensable role in creating truly intelligent and resilient oil field operations.

11.7.5 Limitations and Considerations

- **Regulatory Compliance**: Drone operations must comply with local aviation and safety regulations.
- **Data Overload**: High volumes of data require robust storage, processing, and analysis pipelines.
- **Battery Life and Range**: Operational limits can constrain long missions without support infrastructure.
- **Weather Dependency**: Wind, rain, and dust can hinder drone flights, particularly offshore.

Robots and drones are no longer experimental in DOFs—they are essential. As DOFs mature, robots and drones have been evolving from being inspection aids to autonomous agents of the DOF ecosystem. Their ability to reduce risk, enhance insight, and improve environmental compliance positions them as indispensable tools in the intelligent energy infrastructure of the future.

This future is not a matter of "when," but "how fast" it will be reality! The smart and intelligent field is no longer an upgrade of legacy systems—it is a fundamental reimagination of how energy is produced, managed, and used. It combines physical infrastructure with digital intelligence, human ingenuity with artificial learning, and industrial ambition with environmental/sustainability accountability/responsibility.

To lead in this new era, fully integrated energy companies must not only invest in technology but also in people, trust, transparency, and resilience. Those who act technologically, ethically, and collaboratively will shape not just the future of oil and gas but the future of energy itself (fossil fuels, renewables, hydrogen, and nuclear).

In the next section (appendices), we will examine real-world large-scale DOF implementation where these principles have been applied successfully, highlighting lessons learned and measurable results across diverse environments and field types.

References

1 Alnuaim S. DOF and reservoir monitoring course notes and presentations. KFUPM; 2020–2025.

2 Alnuaim S. Personal technical insights and field experience (35+ years) in DOF, Smart Wells, SCADA, and reservoir monitoring. Saudi Aramco; 1987–2022.

3 SPE. The Digital Transformation Initiative: Autonomous Operations in Oil and Gas. SPE Report; 2019.

4 Shell International Exploration & Production. Smart Fields: From Decision Support to Autonomous Operations. SPE Distinguished Lecture; 2018.

5 McKinsey & Company. The Future of Autonomous Operations in Oil and Gas. McKinsey Energy Insights; 2020.

6 World Economic Forum. Digital Transformation Initiative: Oil and Gas Industry. Geneva: WEF; 2017.

7 Bello O, Teodoriu C, Yaqoob T, et al. Application of artificial intelligence techniques in drilling system design and operations: A state of the art review. J. Artif. Intell. Soft. Comput. Res. 2015;5(2):121–139.

8 SPE. Machine Learning in Upstream Oil and Gas: Applications and Challenges. SPE White Paper; 2020.

9 Russell S, Norvig P. Artificial Intelligence: A Modern Approach. 4th ed. Pearson; 2021.

10 Goodfellow I, Bengio Y, Courville A. Deep Learning. MIT Press; 2016.

11 Grieves M, Vickers J. Digital twin: Mitigating unpredictable, undesirable emergent behavior in complex systems. In: Transdisciplinary Perspectives on Complex Systems. Springer; 2017.

12 SPE. Digital Twins in Oil and Gas: Concepts, Applications, and Value. SPE Technical Report; 2021.

13 Siemens Energy. Industrial Digital Twins for Oil and Gas Operations. Siemens White Paper; 2020.

14 Kritzinger W, Karner M, Traar G, et al. Digital twin in manufacturing: A categorical literature review. IFAC-PapersOnLine. 2018;51(11):1016–1022.

15 IPIECA. Digitalization and the Energy Transition. London: IPIECA; 2021.

16 IEA. Net Zero by 2050: A Roadmap for the Global Energy Sector. Paris: International Energy Agency; 2021.

17 World Bank. Carbon Pricing Dashboard. World Bank Group, Washington DC; 2022.

18 ISO. ISO 14064-1: Greenhouse Gases – Quantification and Reporting. Geneva: ISO; 2018.

19 IEA. World Energy Outlook. Paris: International Energy Agency; 2022.

20 IRENA. Renewable Energy in the Oil and Gas Industry. Abu Dhabi: IRENA; 2021.

21 SPE. Electrification and Renewable Integration in Upstream Operations. SPE Technical Paper; 2021.

22 Preskill J. Quantum computing in the NISQ era and beyond. Quantum 2018;2:79.

23 McArdle S, Endo S, Aspuru-Guzik A, et al. Quantum computational chemistry. Rev. Mod. Phys. 2020;92(1):015003.

24 IBM Research. Quantum Computing for Industrial Optimization. IBM White Paper; 2021.

25 SPE. Robotics and Automation in Oil and Gas Operations. SPE Report; 2020.

26 IFR (International Federation of Robotics). World Robotics Report. International Federation of Robotics (IFR), Frankfurt; 2022.

27 Bogue R. Robots in the oil and gas industries. Ind. Robot. 2016;43(6):593–597.

28 FAA. Unmanned Aircraft Systems (UAS) in Industrial Operations. FAA Advisory Circular; 2020.

29 EASA. Guidelines on Drone Operations in Industrial Environments. European Union Aviation Safety Agency (EASA), Cologne; 2021.

30 SPE. Use of UAVs for Inspection and Emissions Monitoring in Oil and Gas. SPE Technical Paper; 2019.
31 Roth J, Bogue R. The use of drones in oil and gas operations. Ind. Robot. 2020;47(3):395–399.
32 NIST. Cybersecurity Framework for Critical Infrastructure. NIST Special Publication; 2018.
33 OECD. Artificial Intelligence and Data Governance. Paris: OECD; 2021.

A

Typical Large-scale DOF Implementations

A.1 DOF Journey and Key Projects

Among the world's leading fully integrated energy companies, major NOCs and IOCs stand out for their scale, technical depth, and ambition in embracing advanced technology, including digital and intelligent transformation. With the world's largest conventional and unconventional oil/gas reserves and expanding upstream/midstream and downstream operations, these companies recognized that digitization was not merely a tool for efficiency but rather a strategic goal to maintain their everlasting global leadership and optimize resource recovery in a cost-conscious and environmentally constrained way. Major IOC and NOC DOF programs, often referred to as the intelligent field (i-Field), have been cornerstones of their unmatched upstream/midstream/downstream transformation, resulting in economically feasible hydrocarbon development cost ($/barrel). These initiatives have evolved since the early 2000s, representing today one of the most sophisticated examples of large-scale smart field implementation in the global energy sector. These DOF strategies preserved the companies' relatively economically feasible development cost advantage and enhanced the high-tech resilience that kept them the most reliable energy providers in the world [1–12].

A.2 Origins and Strategic Objectives

Most of these projects were launched in the early 2000s in response to the rising operational complexity and growing need to maximize the value and reduce the development cost of their newly developed fields, including offshore, unconventional, and shale. The following are some of the objectives of these DOF projects:

- Enhance reservoir recovery through real-time surveillance and control.
- Reduce operational costs by optimizing production and maintenance.
- Improve decision-making speed and accuracy across the upstream lifecycle.

- Enable integrated asset management at the well, surface network, field, and all asset levels.
- Extend asset life through predictive performance modeling.
- Significantly reduce the development cost of these assets.
- Significantly improve the well integrity, safety, environment, and health of its assets and employees.
- Significantly improve its overall sustainability performance and emissions control strategies.

From their inception, these initiatives were envisioned not as a technology add-on but as a necessary and must-have transformation of upstream/midstream/downstream development, operations, workflows, culture, and organizational change.

A.3 Technical Architecture and Components

Most of these DOF models were built on an integrated, hierarchical digital infrastructure designed to collect, transmit, analyze, and act on massive streams of huge data in size, frequency, and types. Key components include the following:

- Downhole sensors and surface instrumentation for continuous pressure, temperature, flow rates, and water cut monitoring.
- In certain wells, fiber-optic monitoring systems for distributed temperature sensing (DTS) and distributed acoustic sensing (DAS).
- SCADA systems for real-time data acquisition and remote control.
- High-performance computing environments for reservoir simulation and digital twin modeling.
- Integrated operation centers (IOCs) that centralize decision-making across multidisciplines.
- Digital twins of wells, facilities, and entire reservoirs to simulate performance in real time.

These components are supported by skilled human resources and a robust data integration layer that feeds into AI and ML models, enabling real-time production optimization, field monitoring, reservoir management, and anomaly detection.

A.4 Key Technologies and Innovations

Major IOCs and NOCs have developed and deployed numerous DOF proprietary and customized technologies as part of their DOF initiatives, including but not limited to:

- 4D reservoir management systems, integrating time-lapse seismic, well log, and production data to track fluid movements.

- Automated well testing and optimization, reducing the need for manual interventions.
- AI-driven water flood management, particularly critical in maintaining pressure in mature carbonate fields.
- Field-wide surveillance dashboards that give management/engineers a live "digital cockpit" view of well integrity, well productivity, and field performance.
- Remote operations and exception-based surveillance, allowing engineers to focus only on wells and assets showing anomalies.
- ICV optimization, ESP energy optimization, ESP tripping, and failure prediction.
- AI-enabled workflows to predict failures, maintenance needs, and surface/subsurface debottlenecks.

A.5 Measurable Impact ($$$$$)

Several public publications and reports indicated that these DOF investments have yielded substantial returns across technical, operational, and financial dimensions, such as:

- Recovery factor improvements in key fields through enhanced reservoir management.
- Reduction in well intervention costs through condition-based maintenance.
- Enhanced safety due to decreased exposure of field personnel to hazardous environments.
- Carbon intensity reductions, as intelligent energy use has led to better fuel and flare management.
- Production stability, even during external shocks such as market disruptions or logistical constraints.

Moreover, the integration of digital intelligence into asset management has shortened decision cycles from days to hours or even minutes, leading to a historical digitization leadership and oil and gas intelligence added value, helping its major companies such as Saudi Aramco achieving its motto as the most reliable energy provider in the world.

A.6 Organizational and Cultural Transformation

One of the less visible—but equally critical—dimensions of these DOF initiatives has been its impact on these companies' workforce and organizational structure:

- Multidisciplinary collaboration is now standard, with reservoir engineers, geologists, production technologists, and IT specialists working together in real time.

- Upskilling programs have been implemented to train petroleum engineers in data science, AI, and advanced analytics.
- Decision-making hierarchies have flattened, as more decisions are driven by data availability.

Several of these companies have also created centers of excellence in producing and engineering operating areas to serve as incubators for innovation in digital oil field technologies and to support standardization across different assets.

A.7 Challenges and Lessons Learned

Despite their successes and reported high ROI, major IOC and NOC DOF journeys have faced challenges that were effectively and technically addressed:

- Legacy system integration was initially complex, requiring custom middleware and data harmonization strategies.
- Cybersecurity became a top priority as more field equipment was connected to central networks.
- Cultural resistance, which was successfully addressed through planned change management strategies and internal advocacy.
- Data quality and governance were critical early hurdles, necessitating rigorous metadata standards and QA processes.

These companies' responses were to treat the DOF not as a project with a start and end but as an evolving platform, one that is continuously refined based on operational feedback and technological advances. They have continued to invest in the next generation of smart and intelligent field technologies, with a future vision that includes:

- AI-driven autonomous field control, where systems can self-adjust with minimal human input.
- Real-time environmental reporting, using digital sensors to monitor emissions, spills, and flaring in near real time.
- Expanded use of digital twins for end-to-end asset lifecycle management, from drilling to decommissioning.
- Quantum gigascale reservoir simulation and edge-based AI for extremely high recovery and low development cost targets.

As these companies align with global energy transition and emission reduction stretched goals, their DOF architectures are becoming a strategic enabler not only for economic performance but also for sustainability leadership. Their countless digital oil field initiatives are becoming benchmarks for how large-scale, data-intensive, and complex oil and gas operations can be transformed through digital intelligence. This exemplified how the convergence of technology, talent, and strategy can unlock new value from legacy assets while preparing for a lower-carbon, data-driven future. The reported companies' successes demonstrated that the digital oil field is not just a vision; it is a proven business model for the future next generation of upstream/ midstream and downstream operations. All of these companies' DOF initiatives demonstrated significant measurable gains in production efficiency, uptime, and reservoir recovery, validating the concept of digital transformation, resulting in global energy security (supplies), a stable global oil market, and acceptable global energy prices. The level of DOF success varies from one company to another based on its own specific design and implementation. The higher the DOF Readiness Index (DOFRI, Appendix C) and the more smart DOF workflows covering upstream, midstream, and downstream operations in a company, the higher the DOF success. For more DOF implementation details and design, it is highly recommended to visit their website and publicly published technical papers and reports.

References

1 Alnuaim Sami. DOF and Reservoir Monitoring Course Notes & Presentations. KFUPM; 2025.

2 Alnuaim Sami. Personal Technical Insights and Field Experience (35+ years) in DOF, Smart Wells, SCADA, and reservoir monitoring. Saudi Aramco; 1987–2022.

3 Al-Arnaout IH, Al-Dirweesh SM, Al-Zahrani RM. Production engineering experience with the first intelligent field implemented in Saudi Aramco at Haradh-III. SPE Prod. Oper. 2008;23(4):447–456. SPE-112216.

4 Al-Dhubaib TA, Barghouty MF, Issaka MB, et al. Mubarak S, Dowais AH. Saudi Aramco intelligent field development approach: Building the surveillance layer. In: SPE Intelligent Energy Conference; 2008. SPE-112106.

5 Al-Khamis MN, Al-Hamzani HJ, Al-Adel MF. Revamping mature fields using intelligent field technologies. In: SPE Annual Technical Conference and Exhibition; 2009. SPE-123540.

6 Al-Mulhim WA, Faddagh HA, Shehab MA, Shamrani SS, et al. Mega intelligent field application in the world's largest oil increment development. In: SPE Intelligent Energy Conference; 2010. SPE-128837.

7 Al-Kahtani A, Nasr T. Intelligent fields: Saudi Aramco's approach to digital oil field development. J. Pet. Technol. 2019;71(6):34–40.

8 Saudi Aramco. Technology and Innovation for a Sustainable Future. Dhahran: Saudi Aramco; 2022.

9 Theenergyyear.com. Tech transforms Saudi energy: Smart Fields In Action – Khurais; June 2025.

10 Abdelghani H. Saudi Aramco discusses new technology strategy. SPE JPT; June 2013.

11 Saudi Aramco. AI and big data: How Aramco is unlocking the potential of artificial intelligence (AI) and big data to help boost efficiency in our operations; 2020. www.saudiaramco.com.sa.

12 Wallace R. Saudi Aramco, explores the digital technologies the company are deploying in the pursuit of sustainability, efficiency and safety. Oil Field Technology, May 2020.

References

B

DOF AI Case Study: Electrical Submersible Pump (ESP) Failure Prediction

B.1 Introduction

This appendix describes a generic AI example of how to harness digital oil field data streams and artificial intelligence to predict failures of electrical submersible pumps (ESPs). Figures, tables, and codes were generically AI generated for illustration purposes [1–5].

Digital oil field and ESP monitoring platforms integrate multi-ESP, supervisory control and data acquisition, and Plant Information (PI) streams (downhole gauges, variable speed drive [VSD], and motor control center [MCC], telemetry, well tests, and event logs) for real-time analytics. ESPs lift high-volume wells but are subject to electrical, mechanical, and flow assurance failures. Predicting failures days to weeks ahead reduces deferred production and development cost ($/barrel) due to ESP failure prevention and rig time overhead cost for ESP replacement. The AI solution may also recommend changing certain pump operating parameters that would prevent the pump from failure and trip, extending its run life.

B.2 Data Sources and Common Failure Modes

- **Surface and Electrical**: VSD frequency, voltage, current, power factor, total harmonic distortion (THD)/harmonics, and trips (under/overcurrent and ground fault).
- **Downhole**: Intake/discharge pressure and temperature, motor oil temperature, and vibration (if available).
- **Production Context**: Flow rate (oil/water/gas), gas–oil ratio, choke settings, well tests, workovers, and power supply distribution events.
- **Failure Modes**: Electrical faults, motor degradation, cable/connection issues, gas lock, solids/scale, seal/bearing wear, intake starvation/low inflow, pump hydraulics off curve. Table B.1 includes potential signal features and their indicative explanation.

Table B.1 Potential typical ESP signal features and their indicative.

Signal/Feature	Indicative of (Examples)
Rising current + dropping intake pressure	Inflow decline, intake starvation, scaling
Pressure factor (PF) drop + THD rise	Efficiency problem + electrical power quality issues; impending trips
Vibration root mean square (RMS)/kurtosis ↑	Mechanical wear (bearings/seal), true electric stress level
Discharge-intake ΔP off pump curve	Hydraulic degradation/gas interference
Step change in frequency with unstable pressure	Controller hunting/off-design operation

B.3 Problem Framing

- **Binary Early Warning Classification**: Will an ESP fail in the next X days? Labels from workover logs are aligned to time windows.
- **Remaining Useful Life (RUL)/Survival Analysis**: Learn time-to-event from censored run-life data.
- **Unsupervised Anomaly Detection**: Learn a healthy baseline and flag deviations (useful for new wells).

Feature Engineering from Real-time Streams: Compute rolling statistics (mean, std, min/max), slopes/accelerations, change-points, and spectral features from electrical signals. Derive physics-aware features: dynamic head, pump efficiency versus catalog curves, gas fraction proxies (PF↓ and current↑), and temperature deltas. Aggregate events (trips, restart counts) over windows (e.g., 24h/7d/30d).

Algorithms: Gradient-boosted trees for tabular features; sequence models (transformers) for multivariate time series; autoencoders for anomalies; survival models for RUL. Hybrid physics-informed ML constrains models to pump physics.

B.4 Evaluation and Business Value

Use precision-recall AUC (area under the ROC—receiving operating characteristics) on imbalanced data, early warning utility with lead-time bins, and cost–benefit analysis (deferred production vs. intervention cost). Track false alarm rates per well per month and the distribution of lead times.

B.5 Real-time Deployment Pattern

Stream signals → feature service → model server → alerting (tickets/dashboards). Edge inferencing at the remote terminal unit/motor control unit reduces latency; cloud retraining handles fleet learning. Monitor drift, recalibrate thresholds, and provide engineer feedback loops.

B.6 Examples and Illustrations

Figure B.1 shows the multivariate signals prefailure drift (shaded).

Figure B.2 shows the system-generated predicted ESP survival probability versus time.

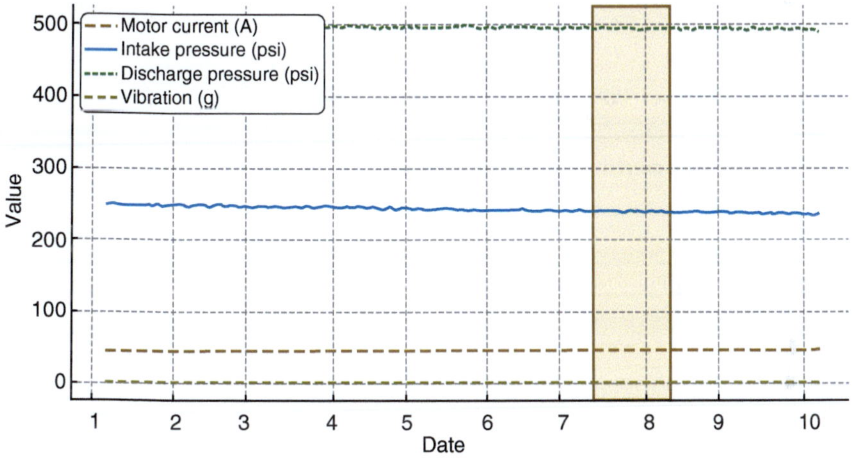

Figure B.1 Real-time ESP signals with prefailure window (illustrative).

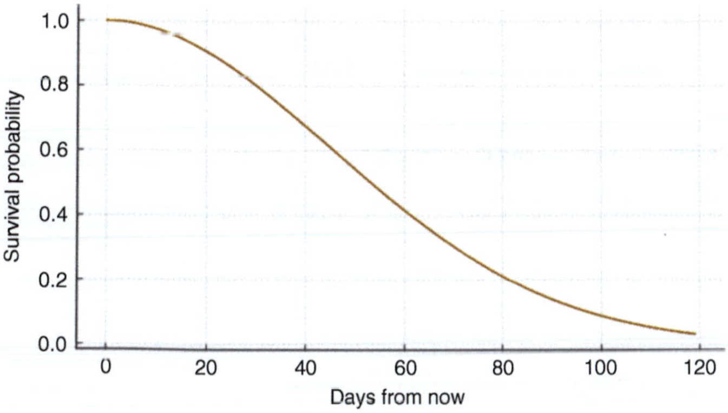

Figure B.2 ESP predicted survival probability versus time (illustrative).

Figure B.3 shows the motor current signature analysis (FFT), revealing sidebands near 60 Hz and harmonics.

Figure B.4 shows the ESP hypothetical feature importance from a gradient-boosted model.

Figure B.5 shows a typical generic AI-generated minimal end-to-end workflow (pseudocode).

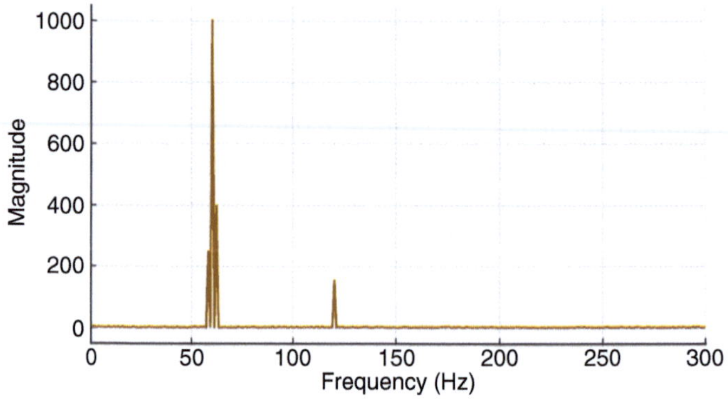

Figure B.3 ESP motor current spectrum (illustrative).

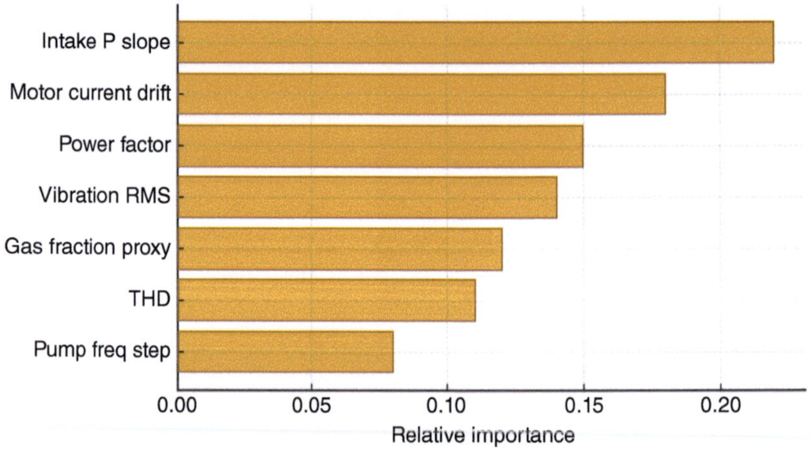

Figure B.4 ESP model features importance (illustrative).

```
# Ingest & resample X = resample_streams(scada_signals, '5min') # Label for early warning within 14
days y = label_failures(workover_log, horizon_days=14) # Feature engineering (windowed stats +
physics-derived) F = make_features(X, windows=[1h, 6h, 24h, 7d], physics=True) # Train & calibrate
model = GradientBoosting().fit(F.train, y.train) thresh = calibrate_threshold(model, F.val, y.val,
target_precision=0.8) # Online inference for batch in stream(F.live): p =
model.predict_proba(batch) if p > thresh and lead_time(batch) >= 5*d: alert(well_id, p,
top_features(model, batch))
```

Generic: For Illustration Purposes

Figure B.5 ESP AI-generated minimal end-to-end workflow (illustrative pseudocode).

B.7 Implementation Checklist

- Define failure taxonomy and label rules; handle censored runs for survival models.
- Build a feature store with rolling windows, slopes, events, and physics-aware metrics.
- Choose framing: 7–30 days horizon classification and/or RUL via survival models.
- Address imbalance and drift; use cross-field validation across wells/fields.
- Embed explainability (feature attributions) in alerts; enable engineer feedback loops.

References

1 Alnuaim Sami. DOF and Reservoir Monitoring Course Notes & Presentations. KFUPM; 2025.

2 Alnuaim Sami. Personal Technical Insights and Field Experience (35+ years) in DOF, Smart Wells, SCADA, and reservoir monitoring. Saudi Aramco; 1987–2022.

3 Abdelkerim A, Abdelkerim I, Alseedi K, Al-Radhi M. Application of AI foresight model in ADNOC offshore for preventing premature ESP failure and extending overall run life; 2025. SPE-224462-MS.

4 Al-Ballam S. A hybrid physical and machine learning model to diagnose failures in electrical submersible pump; 2023. SPE/IADC-214632-MS.

5 Shejuti S, Yocheved G, Thomas AW, Akbari B, Furlong ER, Hughes B. Case study: Predicting electrical submersible pump failures using artificial intelligence and physics-based hybrid models; 2023. SPE-214462-MS.

C

DOF Readiness Index—A New Proposed Framework for Smart Field Transformation

C.1 Introduction

Digital transformation in the oil and gas sector is no longer a future trend—it is a strategic imperative. As companies pursue automation, real-time analytics, and remote operations, a structured approach is needed to evaluate how ready a field or organization is to adopt and benefit from digital oil field (DOF) technologies. This chapter introduces the digital oil field readiness index (DOFRI) concept and proposes a new comprehensive framework that measures oil fields' digital maturity across multiple dimensions [1–8].

C.2 Why a Readiness Index?

Although DOF initiatives aim to:

- Increase reserves and production,
- Reduce development and operational costs, and
- Improve HSE and sustainability,

these outcomes are results of digital maturity—not inputs. A readiness index evaluates whether the technological, operational, and cultural foundations are in place to achieve such outcomes.

C.3 DOF Readiness Dimensions

The proposed DOFRI evaluates the following seven critical domains (Table C.1):

Table C.1 The proposed seven domains for DOFRI.

Dimension	Focus Areas
1. Technology and infrastructure	IoT, SCADA, edge/cloud computing, digital twins
2. Data management and analytics	Real-time data, big data tools, cybersecurity, interoperability
3. Automation and AI integration	Robotics, predictive maintenance, automated workflows
4. Connectivity and communication	5G, satellite links, network redundancy
5. Workforce and digital culture	Digital literacy, AR/VR tools, adaptability to change
6. Operational efficiency and sustainability	NPT reduction, emissions monitoring, energy optimization
7. Regulatory and cybersecurity compliance	ISO/NIST compliance, data governance, risk frameworks

C.4 The Proposed DOFRI Assessment Framework

Each dimension is broken down into specific indicators and weighted based on strategic importance. The following is the proposed weight per dimension:

Example of Weights (out of 100%)

- Technology and infrastructure—20%
- Data management and analytics—15%
- Automation and AI—15%
- Connectivity and communication—10%
- Workforce and digital culture—15%
- Operational efficiency and sustainability—15%
- Regulatory and security compliance—10%

C.5 DOFRI Scoring Methodology

Each indicator is scored from 1 to 5, based on maturity level (Table C.2).

C.6 Sample Indicators by Dimension

Technology and Infrastructure (20%: Each Should Be Evaluated from 1 to 5)

- IoT sensors and SCADA deployment
- Edge computing and cloud adoption
- AI/ML for predictive maintenance

Table C.2 Assessment of each subdimension.

Score	Definition
1	Basic: No implementation
2	Emerging: Initial steps taken
3	Developing: Moderate adoption
4	Mature: Integrated digital tools in place
5	Advanced: Fully autonomous, AI-driven

- Digital twin implementation
- Blockchain for data security and transactions

Data Management and Analytics (15%: Each Should Be Evaluated from 1 to 5)

- Real-time data acquisition and storage
- Big data analytics for decision-making
- Cybersecurity frameworks
- Data integration and interoperability

Automation and AI Integration (15%: Each Should Be Evaluated from 1 to 5)

- Robotics and autonomous operations
- AI-driven predictive maintenance
- Automated workflows for drilling & production
- Digital work instructions and AR/VR use

Connectivity and Communication (10%: Each Should Be Evaluated from 1 to 5)

- High-speed connectivity (5G, fiber, satellite)
- Remote operations capabilities
- Network reliability and redundancy
- Cloud-based collaboration tools

Workforce Readiness (15%: Each Should Be Evaluated from 1 to 5)

- Workforce readiness and digital culture
- Employee digital literacy and training programs
- Change management strategies
- Digital leadership and innovation mindset
- AI-driven workforce augmentation

Operational Efficiency and Sustainability (15%: Each Should Be Evaluated from 1 to 5)

- Reduction in nonproductive time
- Energy efficiency improvements
- Emissions monitoring via digital tools
- AI-driven reservoir management

Regulatory and Security Compliance (10%: Each Should Be Evaluated from 1 to 5)

- Industry standards compliance (ISO 27001, ISO 50001, NIST)
- Cybersecurity risk assessment
- Environmental and safety digital tracking
- AI and data governance frameworks

Final Readiness Index is Calculated as

Each subtitle under a specific dimension is assessed by a DOF expert from 1 to 5 as per Table C.2.

C.6.1 Example

Technology and Infrastructure

- IoT sensors and SCADA deployment — 5: Advanced
- Edge computing and cloud adoption — 4: Mature
- AI/ML for predictive maintenance — 5: Advanced
- Digital twin implementation — 5: Advanced
- Blockchain for data security and transactions — 1: Basic

Technology and infrastructure dimension composite score = (5+4+5+5+1)/5=4

The same method is used to assess the other six dimensions.

Let's assume the averages of dimensions 1–7 are assessed by a DOF consultant as D1–D7. Then the final DOFRI can be calculated as the following equation:

$$\begin{aligned}\text{DOFRI} = 100 * [&(D_1 * 0.2) + (D_2 * 0.15) + (D_3 * 0.15) + (D_4 * 0.1) \\ &+ (D_5 * 0.15) + (D_6 * 0.15) + (D_7 * 0.1)]/5\end{aligned} \tag{C.1}$$

Table C.3 includes the proposed DOFRI levels.

The DOFRI offers a structured, quantitative tool (enable) to assess and enhance an organization's preparedness for harvesting its digital transformation investment. By evaluating key technological, human, and operational domains, the index helps

Table C.3 Proposed DOFRI levels.

Index (%)	Interpretation
0–30%	Low—manual processes dominate
31–60%	Moderate—partial digital adoption
61–85%	High—significant digital infrastructure in place
86–100%	Advanced—fully digital and autonomous operation

companies align their digital strategies with practical execution, ensuring a measurable path to higher performance and resilience in the energy sector. Once readiness is assessed, companies can further relate it to other value outcomes depending on their heavy use of DOF smart systems and workflows, fulfilling the following three main DOF strategic objectives:

1. **Increased Production/Reserves**
   - **Exploration**: Faster cycle times from exploration, drill to first oil.
   - **Production**: Optimum high hydrocarbon and low undesired fluids production.
   - **Reservoir Management**: Real-time optimization and reserves/recovery assessment.
2. **Reduced CAPEX/OPEX/Development Cost**
   - **Production and Workover**: Reduction of manual/workover interventions and operation downtown.
   - **Producing**: Reduction in energy consumption, increase in process efficiency, and upstream operations (oil and gas energy supply) reliability.
   - **Maintenance**: Significant improvement in facilities maintenance (cycle, cost, and quality) resulting into significant reduction in operational cost and facilities' downtime.
3. **Enhanced HSE and Sustainability**
   - **Safety**: Reduction in workforce hazardous exposure and travel.
   - **Emission**: Improvement in emission control, tracking, reporting, and compliance.
   - **Security**: Improvement in physical assets' security and ultimate energy security.

The overall DOF evaluation and return on investment (ROI) will be the combination of both the DOFRI (the enabler) and the overall increase in recovery/production, reduction in operational cost (CAPEX/OPEX), and improvement in ESG performance, attributed mainly to the DOF implementation strategy. The higher the number of smart DOF workflows' implementations and the higher the KPI of the above three DOF strategic objectives, the higher the DOF ROI.

Table C.4 Typical world-class field A DOF assessment.

DOF Dimension	Score (1–5)	Notes
Technology and infrastructure	4.5	IoT sensors, SCADA, AI/ML, digital twins, blockchain
Data management and analytics	5	Real-time data, big data, data availability, data accuracy, cybersecurity
Automation and AI integration	5	Robotics, AI-driven workflows, AR/VR
Connectivity and communication	5	5G/fiber, cloud collaboration, satellite
Workforce and digital culture	4.5	Upskilling, digital leadership, innovation
Operational efficiency and sustainability	4.5	APC, energy optimization, emissions tracking
Regulatory and cybersecurity compliance	4	ISO standards, AI and data governance, environmental compliance

C.7 Example of a Typical DOF Readiness Index Assessment

Field A, a typical world-class DOF, was assessed by a DOF consultant as per Table C.4.

$$\text{DOFRI} = 100 * [(4.5 * 0.2) + (5 * 0.15) + (5 * 0.15) + (5 * 0.1) + (4.5 * 0.15) + (4.5 * 0.15) + (4 * 0.1)]/5 = 93\%$$

Final Readiness Score: **93%**, which is considered an advanced, fully world-class digital and automated smart oil field. This high score is considered to be an enabler for achieving the DOF strategic objectives of increasing production/reserves, reducing CAPEX/OPEX, and enhancing HSE and sustainability performances. The added value of such a technology on the overall field development can be further $ quantified, reflecting a reduction in the development per barrel ($/barrel).

References

1 Alnuaim Sami. DOF and Reservoir Monitoring Course Notes & Presentations. KFUPM; 2025.

2 Alnuaim Sami. Personal Technical Insights and Field Experience (35+ years) in DOF, Smart Wells, SCADA, and reservoir monitoring. Saudi Aramco; 1987–2022.

3 Digital oilfields: The Next Generation of Intelligent Operations. Deloitte Insights; 2022.
4 Accenture. Digital Maturity in Oil and Gas. Accenture Energy Report; 2022.
5 Cann G, Goydan R. Bits, Bytes, and Barrels: The Digital Transformation of Oil and Gas. Houston: Gulf Publishing; 2019.
6 Emerging Technologies in Industrial Digitalization. Stamford: Gartner Research; 2022.
7 Khurais Intelligent Field & Sustainability Performance, Saudi Aramco. Khurais: Tthe Smart Field of the Future. Dhahran: Saudi Aramco; 2020.
8 Al Jadidi I, AL-Saqri KA, Al Busaidi HM, Al-Mujaini AS, Al Ghafri RS, Alsharji H, Al-Ghadani A. Petroleum Development Oman, Muscat, Muscat Governorate, Sultanate of Oman. 2018.

Glossary of Terms

A

AI (Artificial Intelligence) Computational systems that perform tasks normally requiring human intelligence, such as learning, pattern recognition, and predictive analytics. Used in DOF for optimization, anomaly detection, and decision support.

API (American Petroleum Institute) Organization that establishes widely adopted technical standards for oil and gas equipment and operations.

Asset Management Integrated process of monitoring, maintaining, and optimizing oil field assets throughout their lifecycle to maximize value and reliability.

B

Big Data Extremely large datasets derived from sensors, logs, and models across surface and subsurface systems, analyzed for insights to improve operational decisions.

Brown Field Mature or existing oil and gas field that relies on conventional analog or partially digital operations, often targeted for DOF retrofitting.

C

CCE (Carbon Circular Economy) Framework emphasizing reuse, recycling, and conversion of carbon emissions into valuable products to achieve net-zero operations.

Cloud Computing On-demand network access to scalable computing resources enabling real-time data storage, analytics, and collaboration across global operations.

Control System Combination of devices (DCS, PLC, SCADA) managing and automating field processes.

Cybersecurity Protection of digital oil field systems and networks from unauthorized access or cyber threats.

D

DCS (Distributed Control System) Automation architecture that monitors and controls processes distributed throughout a facility or field.

Digital Oil Field (DOF) Integrated, data-driven operational environment combining real-time monitoring, analytics, and automation to optimize production, safety, and sustainability.

Digital Twin Virtual replica of a physical asset or process used for simulation, optimization, and predictive analysis.

DTS (Distributed Temperature Sensing) Fiber-optic-based system measuring temperature continuously along a wellbore to detect leaks or monitor flow profiles.

E

Edge Computing Localized data processing at or near field devices to enable real-time control even with limited connectivity.

ESG (Environmental, Social, and Governance) Corporate framework ensuring responsible and sustainable operations in line with environmental and social obligations.

ESP (Electrical Submersible Pump) Downhole artificial lift system that boosts fluid production from reservoirs to the surface.

F

Fiber-Optic Sensor Optical-fiber-based sensor used for distributed temperature or acoustic measurements within the wellbore.

Flow Assurance Discipline ensuring uninterrupted, safe flow of hydrocarbons from reservoir to market through proper design and monitoring.

H

HSE (Health, Safety, and Environment) Critical operational pillar focusing on minimizing risk to personnel, assets, and the environment.

Hydraulic Control Line Tubing line transmitting hydraulic pressure from the surface to operate downhole tools such as inflow control valves or safety valves.

I

ICD (Inflow Control Device) Passive completion device managing inflow along the wellbore to improve sweep efficiency.

ICV (Inflow Control Valve) Actuated valve in smart completions allowing remote zonal flow control to optimize production and delay water/gas breakthrough.

IIoT (Industrial Internet of Things) Network of connected industrial devices enabling real-time data collection and analytics.

Integrated Operations Center (IOC) Centralized hub where multidisciplinary teams monitor, analyze, and control real-time field operations.

ISO (International Organization for Standardization) Global standards body governing equipment, quality, and safety compliance.

L

LTIs (Lost-Time Incidents) HSE metric representing injuries causing lost work time; DOF aims to minimize LTIs via remote and automated operations.

M

ML (Machine Learning) Subfield of AI that trains algorithms to recognize patterns and make data-driven predictions, for example, ESP failure forecasting.

MPFM (Multiphase Flow Meter) Instrument measuring simultaneous flow rates of oil, gas, and water in real time without phase separation.

N

NACE International organization that sets standards for corrosion-control materials and practices in oil and gas environments.

NOC (National Oil Company) State-owned petroleum company managing exploration and production within a country.

NPV (Net Present Value) Financial metric evaluating the profitability of DOF investments through discounted cash flow analysis.

O

OPEX (Operating Expenditure) Day-to-day expenses required to run oil field operations; DOF reduces OPEX via automation and predictive maintenance.

OSC (Oil Services Company) Specialized firm providing technological and field services, such as Schlumberger or Halliburton.

P

PDHMS (Permanent Downhole Measurement System) Permanently installed downhole-sensor system providing continuous pressure and temperature data.

PI System (Plant Information System) Data management platform aggregating time-series operational data for visualization and analytics.

PLC (Programmable Logic Controller) Industrial digital computer controlling process equipment in real time.

Predictive Maintenance Maintenance strategy using data analytics and AI to anticipate and prevent equipment failures before they occur.

Q

Quantum Computing Emerging computational paradigm leveraging quantum mechanics to perform complex simulations, for example, for reservoir modeling.

R

Reservoir Digital Twin Real-time simulation model replicating reservoir dynamics for optimization of production and injection strategies.

RTU (Remote Telemetry Unit) Device that collects sensor data and transmits it to control centers via communication networks.

S

SCADA (Supervisory Control and Data Acquisition) System architecture enabling remote monitoring and control of wells, pipelines, and facilities.

Smart Well Well equipped with intelligent completions and sensors enabling remote monitoring and control of production parameters.

SSSV (Subsurface Safety Valve) Downhole safety valve that automatically shuts off well flow during emergencies.

Sustainability Performance Measurement of environmental and social responsibility outcomes, including emissions and energy efficiency.

T

Telemetry System Communication infrastructure transferring sensor data from field equipment to surface control systems.

Total Field Optimization (TFO) Integrated optimization of reservoir, well, and surface network performance using DOF analytics.

V

VSD (Variable Speed Drive) Electrical device controlling motor speed and torque, allowing ESP flow-rate optimization and energy savings.

W

Well Integrity Assurance that a well's barriers (casing, tubing, packers) prevent uncontrolled fluid migration across formations or to the surface.

Workflow Automation Streamlined digital process linking data inputs to analytical outputs and control actions for efficient decision-making.

Index

Note: Page numbers in **bold** and *italics* refers to tables and figures.